WASTEWATER TREATMENT FUNDAMENTALS I

LIQUID TREATMENT

OPERATOR CERTIFICATION STUDY QUESTIONS

2020

Water Environment Federation
601 Wythe Street
Alexandria, VA 22314–1994 USA
http://www.wef.org

Association of Boards of Certification
2805 SW Snyder Blvd., Suite 535
Ankeny, IA 50023
http://www.abccert.org

IMPORTANT NOTICE

The material presented in this publication has been prepared in accordance with generally recognized engineering principles and practices and is for general information only. This information should not be used without first securing competent advice with respect to its suitability for any general or specific application.

The contents of this publication are not intended to be a standard of the Water Environment Federation (WEF) or the Association of Boards of Certification (ABC) and are not intended for use as a reference in purchase specifications, contracts, regulations, statutes, or any other legal document.

No reference made in this publication to any specific method, product, process, or service constitutes or implies an endorsement, recommendation, or warranty thereof by WEF or ABC.

WEF and ABC make no representation or warranty of any kind, whether expressed or implied, concerning the accuracy, product, or process discussed in this publication and assumes no liability.

Anyone using this information assumes all liability arising from such use, including but not limited to infringement of any patent or patents.

About WEF

The Water Environment Federation (WEF) is a not-for-profit technical and educational organization of 33,000 individual members and 75 affiliated Member Associations representing water quality professionals around the world. Since 1928, WEF and its members have protected public health and the environment. As a global water sector leader, our mission is to connect water professionals; enrich the expertise of water professionals; increase the awareness of the impact and value of water; and provide a platform for water sector innovation. To learn more, visit www.wef.org.

About ABC

The Association of Boards of Certification (ABC) was founded with the mission to advance the quality and integrity of environmental certification programs throughout the world. This charge has held strong through more than 40 years of providing knowledge and resources to nearly 100 certifying authorities representing more than 40 states, 10 Canadian provinces and territories, and several international and tribal programs. ABC believes in certification as a means of promoting public health and the environment while striving to give our members the necessary tools to ensure the knowledge and skills of their operators.

Preface

These more than 800 study questions and answers represent all aspects of liquid treatment processes and help operators prepare for the first three levels of certification examinations and measure operators' knowledge of the basics of liquid treatment as well as critical aspects of biological treatment, nutrient removal, and disinfection.

These questions are also included as "Test Your Knowledge" questions in the *Wastewater Treatment Fundamentals I—Liquid Treatment* training manual and online course developed in collaboration with the Association of Boards of Certification (ABC). This separate study guide is intended for those operators that do not have the opportunity to use the training manual or online course for study.

Prepared by

Sidney Innerebner, Ph.D., P.E., CWP, PO, A Industrial Wastewater and A Municipal Wastewater Operator in Colorado (Certified Water Professional), Grade IV Professional Wastewater Operator, Indigo Water Group, *Contractor*

With assistance and review provided by

Mike Bell, Supervisor, City of Hamilton, Hamilton, Ontario, Canada (Retired) (ABC)

Mark Cherniak

Nathan Coey, Pataskala Ohio Utility Director

Frank J. DeOrio, PO – Sr. Technical Director, O'Brien & Gere, Syracuse, New York (ABC)

Justin Elkins, City of Louisville, Colorado

John J. Fortin, P.E., Hazen and Sawyer, New York, New York

Sarah Galst

Mike Gosselin, PO (ABC)

Paul Krauth, Statepoint Engineering

Shaun Livermore, PO – Operations Manager, Poarch Band of Creek Indians Utility Authority, Atmore, Alabama (ABC)

Jorj Long

Michael P. Lutz, P.E., Principal Engineer, Dewberry Engineers, Inc., Denver, Colorado

Timothy Meloveck, PO, CWP – Utilities Supervisor, Town of Carbondale, Carbondale, Colorado (ABC)

Andy O'Neill, PO – Environment Specialist, Washington State Department of Ecology, Spokane, Washington (ABC)

Stacy J. Passaro, P.E., Passaro Engineering, LLC, Mount Airy, Maryland

Paul Pitt

Mark Poling

John R. Reynolds, PO – Contract Operator and Instructor, Sooke, British Columbia, Canada (ABC)

Kim Riddell, Alloway, Lima, Ohio

Joel C. Rife, P.E.

LeAnna Risso, PO – Process Control Manager, Clark County Water Reclamation District (ABC)

Kenneth Schnaars, P.E., Brown and Caldwell

Greg Seaman, President, O&M Solutions, LLC, Howell, New Jersey (ABC)

Ron Trygar

Jack Vanderland, Director of Wastewater Operator Training and Assistance Program, Virginia Department of Environmental Quality, Richmond, Virginia (Retired) (ABC)

Rich Weigand

Chris White, P.E., Hazen and Sawyer

Brian Woods, Clifton Sanitation District

Under the Direction of the Wastewater Treatment Fundamentals Steering Group of the **Technical Practice Committee**

Rudy Fernandez
John Hart
Cordell Samuels

Contents

Chapter 9   Nutrient Removal 89

Chapter 10   Disinfection 101

Association of Boards of Certification Formulas
Wastewater Treatment, Collection, Industrial Waste, & Wastewater Laboratory Exams

*Pie wheel format for this equation is shown at the end of the formulas.

$$\text{Alkalinity, mg/L as CaCO}_3 = \frac{(\text{Titrant Volume, mL})(\text{Acid Normality})(50\ 000)}{\text{Sample Volume, mL}}$$

$$\text{Amps} = \frac{\text{Volts}}{\text{Ohms}}$$

Area of Circle* = $(0.785)(\text{Diameter}^2)$

Area of Circle = $(3.14)(\text{Radius}^2)$

Area of Cone (lateral area) = $(3.14)(\text{Radius})\sqrt{\text{Radius}^2 + \text{Height}^2}$

Area of Cone (total surface area) = $(3.14)(\text{Radius})(\text{Radius} + \sqrt{\text{Radius}^2 + \text{Height}^2})$

Area of Cylinder (total exterior surface area) = [End #1 SA] + [End #2 SA] + [(3.14)(Diameter)(Height or Depth)]
   Where SA = surface area

Area of Rectangle* = (Length)(Width)

$$\text{Area of Right Triangle*} = \frac{(\text{Base})(\text{Height})}{2}$$

$$\text{Average (arithmetic mean)} = \frac{\text{Sum of All Terms}}{\text{Number of Terms}}$$

Average (geometric mean) = $[(X_1)(X_2)(X_3)(X_4)(X_n)]^{1/n}$ The *nth* root of the product of *n* numbers

$$\text{Biochemical Oxygen Demand (seeded), mg/L} = \frac{[(\text{Initial DO, mg/L}) - (\text{Final DO, mg/L}) - (\text{Seed Correction, mg/L})][300\ \text{mL}]}{\text{Sample Volume, mL}}$$

$$\text{Biochemical Oxygen Demand (unseeded), mg/L} = \frac{[(\text{Initial DO, mg/L}) - (\text{Final DO, mg/L})][300\ \text{mL}]}{\text{Sample Volume, mL}}$$

Blending or Three Normal Equation = $(C_1 \times V_1) + (C_2 \times V_2) = (C_3 \times V_3)$ *Where $V_1 + V_2 = V_3$; C = concentration, V = volume or flow; Concentration units must match; Volume units must match*

$$\text{\# CFU/100 mL} = \frac{[(\text{\# of Colonies on Plate})(100)]}{\text{Sample Volume, mL}}$$

$$\text{Chemical Feed Pump Setting, \% Stroke} = \frac{\text{Desired Flow}}{\text{Maximum Flow}} \times 100\%$$

$$\text{Chemical Feed Pump Setting, mL/min} = \frac{(\text{Flow, mgd})(\text{Dose, mg/L})(3.785\ \text{L/gal})(1\ 000\ 000\ \text{gal/mil. gal})}{(\text{Feed Chemical Density, mg/mL})(\text{Active Chemical, \% express as a decimal})(1440\ \text{min/d})}$$

$$\text{Chemical Feed Pump Setting, mL/min} = \frac{(\text{Flow, m}^3/\text{d})(\text{Dose, mg/L})}{(\text{Feed Chemical Density, g/cm}^3)(\text{Active Chemical, \% express as a decimal})(1440\ \text{min/d})}$$

Circumference of Circle = (3.14)(Diameter)

$$\text{Composite Sample Single Portion} = \frac{(\text{Instantaneous Flow})(\text{Total Sample Volume})}{(\text{Number of Portions})(\text{Average Flow})}$$

$$\text{Cycle Time, min} = \frac{\text{Storage Volume, gal}}{(\text{Pump Capacity, gpm}) - (\text{Wet Well Inflow, gpm})}$$

$$\text{Cycle Time, min} = \frac{\text{Storage Volume, m}^3}{(\text{Pump Capacity, m}^3/\text{min}) - (\text{Wet Well Inflow, m}^3/\text{min})}$$

$$\text{Degrees Celsius} = \frac{(°F - 32)}{1.8}$$

Degrees Fahrenheit = (°C)(1.8) + 32

$$\text{Detention Time} = \frac{\text{Volume}}{\text{Flow}} \qquad \textit{Units must be compatible}$$

Dilution or Two Normal Equation = $(C_1 \times V_1) = (C_2 \times V_2)$ *Where C = Concentration, V = volume or flow; Concentration units must match; Volume units must match*

Electromotive Force, V* = (Current, A)(Resistance, ohm - Ω)

$$\text{Feed Rate, lb/d*} = \frac{(\text{Dosage, mg/L})(\text{Flow, mgd})(8.34 \text{ lb/gal})}{\text{Purity, \% expressed as a decimal}}$$

$$\text{Feed Rate, kg/d*} = \frac{(\text{Dosage, mg/L})(\text{Flow rate, m}^3/\text{d})}{(\text{Purity, \% expressed as a decimal})(1000)}$$

$$\text{Filter Backwash Rate, gpm/sq ft} = \frac{\text{Flow, gpm}}{\text{Filter Area, sq ft}}$$

$$\text{Filter Backwash Rate, L/(m}^2\text{·s)} = \frac{\text{Flow, L/s}}{\text{Filter Area, m}^2}$$

$$\text{Filter Backwash Rise Rate, in./min} = \frac{(\text{Backwash Rate, gpm/sq ft})(12 \text{ in./ft})}{7.48 \text{ gal/cu ft}}$$

$$\text{Filter Backwash Rise Rate, cm/min} = \frac{\text{Water Rise, cm}}{\text{Time, min}}$$

$$\text{Filter Yield, lb/sq ft/hr} = \frac{(\text{Solids Loading, lb/d})(\text{Recovery, \% expressed as a decimal})}{(\text{Filter Operation, hr/d})(\text{Area, sq ft})}$$

$$\text{Filter Yield, kg/m}^2\text{·h} = \frac{(\text{Solids Concentration, \% expressed as a decimal})(\text{Sludge Feed Rate, L/h})(10)}{(\text{Surface Area of Filter, m}^2)}$$

Flowrate, cu ft/sec* = (Area, sq ft)(Velocity, ft/sec)

Flowrate, m³/sec* = (Area, m²)(Velocity, m/s)

$$\text{Food-to-Microorganism Ratio} = \frac{\text{BOD}_5, \text{ lb/d}}{\text{MLVSS, lb}}$$

$$\text{Food-to-Microorganism Ratio} = \frac{\text{BOD}_5, \text{ kg/d}}{\text{MLVSS, kg}}$$

Force, lb* = (Pressure, psi)(Area, sq in.)

Force, newtons* = (Pressure, Pa)(Area, m²)

Hardness, as mg CaCO₃/L = $\dfrac{\text{(Titrant Volume, mL)(1000)}}{\text{Sample Volume, mL}}$ *Only when the titration factor is 1.00 of ethylenediaminetetraacetic acid (EDTA)*

Horsepower, Brake, hp = $\dfrac{\text{(Flow, gpm)(Head, ft)}}{\text{(3960)(Pump Efficiency, \% expressed as a decimal)}}$

Horsepower, Brake, kW = $\dfrac{\text{(9.8)(Flow, m}^3\text{/s)(Head, m)}}{\text{(Pump Efficiency, \% expressed as a decimal)}}$

Horsepower, Motor, hp = $\dfrac{\text{(Flow, gpm)(Head, ft)}}{\text{(3960)(Pump Efficiency, \% expressed as a decimal)(Motor Efficiency, \% expressed as a decimal)}}$

Horsepower, Motor, kW = $\dfrac{\text{(9.8)(Flow, m}^3\text{/s)(Head, m)}}{\text{(Pump Efficiency, \% expressed as a decimal)(Motor Efficiency, \% expressed as a decimal)}}$

Horsepower, Water, hp = $\dfrac{\text{(Flow, gpm)(Head, ft)}}{3960}$

Horsepower, Water, kW = (9.8)(Flow, m³/s)(Head, m)

Hydraulic Loading Rate, gpd/sq ft = $\dfrac{\text{Total Flow Applied, gpd}}{\text{Area, sq ft}}$

Hydraulic Loading Rate, m³/(m²·d) = $\dfrac{\text{Total Flow Applied, m}^3\text{/d}}{\text{Area, m}^2}$

Loading Rate, lb/d* = (Flow, mgd)(Concentration, mg/L)(8.34 lb/gal)

Loading Rate, kg/d* = $\dfrac{\text{(Flow, m}^3\text{/d)(Concentration, mg/L)}}{1000}$

Mass, lb* = (Volume, mil. gal)(Concentration, mg/L)(8.34 lb/gal)

Mass, kg* = $\dfrac{\text{(Volume, m}^3\text{)(Concentration, mg/L)}}{1000}$

Mean Cell Residence Time or Solids Retention Time, days = $\dfrac{\text{(Aeration Tank TSS, lb)} + \text{(Clarifier TSS, lb)}}{\text{(TSS Wasted, lb/d)} + \text{(Effluent TSS, lb/d)}}$

Milliequivalent = (mL)(Normality)

Molarity = $\dfrac{\text{Moles of Solute}}{\text{Liters of Solution}}$

Motor Efficiency, % = $\dfrac{\text{Brake hp}}{\text{Motor hp}} \times 100\%$

Normality = $\dfrac{\text{Number of Equivalent Weights of Solute}}{\text{Liters of Solution}}$

Number of Equivalent Weights = $\dfrac{\text{Total Weight}}{\text{Equivalent Weight}}$

$$\text{Number of Moles} = \frac{\text{Total Weight}}{\text{Molecular Weight}}$$

$$\text{Organic Loading Rate-RBC, lb SBOD}_5/1000 \text{ sq ft/d} = \frac{\text{Organic Load, lb SBOD}_5/\text{d}}{\text{Surface Area of Media, 1000 sq ft}}$$

$$\text{Organic Loading Rate-RBC, kg SBOD}_5/\text{m}^2{\cdot}\text{d} = \frac{\text{Organic Load, kg SBOD}_5/\text{d}}{\text{Surface Area of Media, m}^2}$$

$$\text{Organic Loading Rate-Trickling Filter, lb BOD}_5/1000 \text{ cu ft/d} = \frac{\text{Organic Load, lb BOD}_5/\text{d}}{\text{Volume, 1000 cu ft}}$$

$$\text{Organic Loading Rate-Trickling Filter, kg/m}^3{\cdot}\text{d} = \frac{\text{Organic Load, kg BOD}_5/\text{d}}{\text{Volume, m}^3}$$

$$\text{Oxygen Uptake Rate or Oxygen Consumption Rate, mg/L}{\cdot}\text{min} = \frac{\text{Oxygen Usage, mg/L}}{\text{Time, min}}$$

$$\text{Population Equivalent, Organic} = \frac{(\text{Flow, mgd})(\text{BOD, mg/L})(8.34 \text{ lb/gal})}{0.17 \text{ lb BOD/d/person}}$$

$$\text{Population Equivalent, Organic} = \frac{(\text{Flow, m}^3/\text{d})(\text{BOD, mg/L})}{(1000)(0.077 \text{ kg BOD/d}{\cdot}\text{person})}$$

$$\text{Power, kW} = \frac{(\text{Flow, L/s})(\text{Head, m})(9.8)}{1000}$$

$$\text{Recirculation Ratio-Trickling Filter} = \frac{\text{Recirculated Flow}}{\text{Primary Effluent Flow}}$$

$$\text{Reduction of Volatile Solids, \%} = \left(\frac{\text{VS in} - \text{VS out}}{\text{VS in} - (\text{VS in} \times \text{VS out})}\right) \times 100\% \qquad \textit{All information (In and Out) must be in decimal form}$$

$$\text{Removal, \%} = \left(\frac{\text{In} - \text{Out}}{\text{In}}\right) \times 100\%$$

$$\text{Return Rate, \%} = \frac{\text{Return Flowrate}}{\text{Influent Flowrate}} \times 100\%$$

$$\text{Return Sludge Rate-Solids Balance, mgd} = \frac{(\text{MLSS, mg/L})(\text{Flowrate, mgd})}{(\text{RAS Suspended Solids, mg/L}) - (\text{MLSS, mg/L})}$$

$$\text{Slope, \%} = \frac{\text{Drop or Rise}}{\text{Distance}} \times 100\%$$

$$\text{Sludge Density Index} = \frac{100}{\text{SVI}}$$

$$\text{Sludge Volume Index, mL/g} = \frac{(\text{SSV}_{30}, \text{mL/L})(1000 \text{ mg/g})}{\text{MLSS, mg/L}}$$

$$\text{Solids, mg/L} = \frac{(\text{Dry Solids, g})(1\,000\,000)}{\text{Sample Volume, mL}}$$

$$\text{Solids Capture, \% (Centrifuges)} = \left[\frac{\text{Cake TS, \%}}{\text{Feed Sludge TS, \%}}\right] \times \left[\frac{(\text{Feed Sludge TS, \%}) - (\text{Centrate TSS, \%})}{(\text{Cake TS, \%}) - (\text{Centrate TSS, \%})}\right] \times 100\%$$

$$\text{Solids Concentration, mg/L} = \frac{\text{Weight, mg}}{\text{Volume, L}}$$

$$\text{Solids Loading Rate, lb/sq ft/d} = \frac{\text{Solids Applied, lb/d}}{\text{Surface Area, sq ft}}$$

$$\text{Solids Loading Rate, kg/m}^2\text{·d} = \frac{\text{Solids Applied, kg/d}}{\text{Surface Area, m}^2}$$

Solids Retention Time: *see Mean Cell Residence Time*

$$\text{Specific Gravity} = \frac{\text{Specific Weight of Substance, lb/gal}}{8.34 \text{ lb/gal}}$$

$$\text{Specific Gravity} = \frac{\text{Specific Weight of Substance, kg/L}}{1.0 \text{ kg/L}}$$

$$\text{Specific Oxygen Uptake Rate or Respiration Rate, (mg/g)/h} = \frac{(\text{OUR, mg/L·min})(60 \text{ min})}{(\text{MLVSS, g/L})(1 \text{ h})}$$

$$\text{Surface Loading Rate or Surface Overflow Rate, gpd/sq ft} = \frac{\text{Flow, gpd}}{\text{Area, sq ft}}$$

$$\text{Surface Loading Rate or Surface Overflow Rate, L/m}^2\text{·d} = \frac{\text{Flow, L/d}}{\text{Area, m}^2}$$

$$\text{Total Solids, \%} = \frac{(\text{Dried Weight, g}) - (\text{Tare Weight, g})}{(\text{Wet Weight, g}) - (\text{Tare Weight, g})} \times 100\%$$

$$\text{Velocity, ft/sec} = \frac{\text{Flowrate, cu ft/s}}{\text{Area, sq ft}}$$

$$\text{Velocity, ft/sec} = \frac{\text{Distance, ft}}{\text{Time, sec}}$$

$$\text{Velocity, m/s} = \frac{\text{Flowrate, m}^3/\text{s}}{\text{Area, m}^2}$$

$$\text{Velocity, m/s} = \frac{\text{Distance, m}}{\text{Time, s}}$$

$$\text{Volatile Solids, \%} = \left[\frac{(\text{Dry Solids, g}) - (\text{Fixed Solids, g})}{(\text{Dry Solids, g})}\right] \times 100\%$$

Volume of Cone* $= (1/3)(0.785)(\text{Diameter}^2)(\text{Height})$

Volume of Cylinder* $= (0.785)(\text{Diameter}^2)(\text{Height})$

Volume of Rectangular Tank* $= (\text{Length})(\text{Width})(\text{Height})$

$$\text{Water Use, gpcd} = \frac{\text{Volume of Water Produced, gpd}}{\text{Population}}$$

$$\text{Water Use, L/cap·d} = \frac{\text{Volume of Water Produced, L/d}}{\text{Population}}$$

Watts (AC circuit) = (Volts)(Amps)(Power Factor)

Watts (DC circuit) = (Volts)(Amps)

$$\text{Weir Overflow Rate, gpd/ft} = \frac{\text{Flow, gpd}}{\text{Weir Length, ft}}$$

$$\text{Weir Overflow Rate, L/m·d} = \frac{\text{Flow, L/d}}{\text{Weir Length, m}}$$

$$\text{Wire-to-Water Efficiency, \%} = \frac{\text{Water hp}}{\text{Motor hp}} \times 100\%$$

$$\text{Wire-to-Water Efficiency, \%} = \frac{(\text{Flow, gpm})(\text{Total Dynamic Head, ft})(0.746 \text{ kW/hp})(100\%)}{(3960)(\text{Electrical Demand, kW})}$$

Pie Wheels

- To find the quantity above the horizontal line: multiply the pie wedges below the line together.
- To solve for one of the pie wedges below the horizontal line: cover that pie wedge, then divide the remaining pie wedge(s) into the quantity above the horizontal line.
- Given units must match the units shown in the pie wheel.
- When US and metric units or values differ, the metric is shown in parentheses, e.g. (m²).

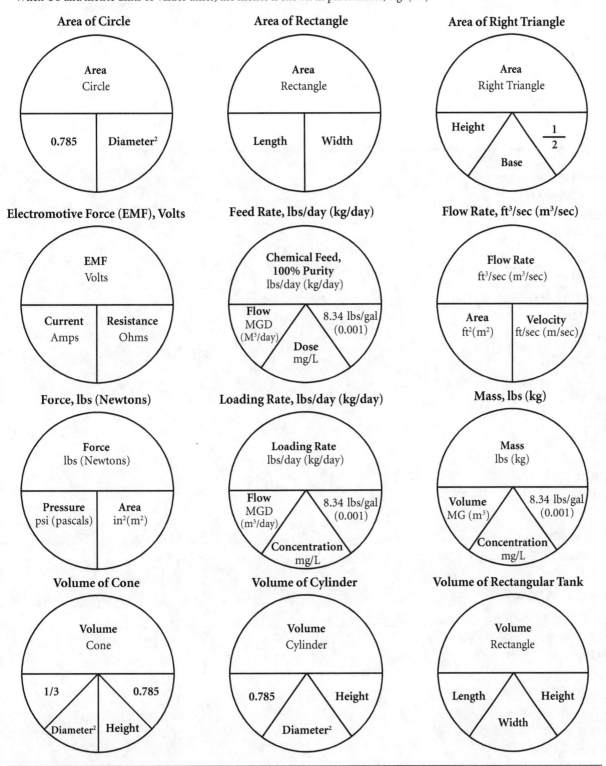

Area of Circle

Area
Circle

0.785 | Diameter²

Area of Rectangle

Area
Rectangle

Length | Width

Area of Right Triangle

Area
Right Triangle

Height | $\frac{1}{2}$
Base

Electromotive Force (EMF), Volts

EMF
Volts

Current
Amps | Resistance
Ohms

Feed Rate, lbs/day (kg/day)

Chemical Feed,
100% Purity
lbs/day (kg/day)

Flow
MGD
(M³/day) | 8.34 lbs/gal
(0.001)
Dose
mg/L

Flow Rate, ft³/sec (m³/sec)

Flow Rate
ft³/sec (m³/sec)

Area
ft²(m²) | Velocity
ft/sec (m/sec)

Force, lbs (Newtons)

Force
lbs (Newtons)

Pressure
psi (pascals) | Area
in²(m²)

Loading Rate, lbs/day (kg/day)

Loading Rate
lbs/day (kg/day)

Flow
MGD
(m³/day) | 8.34 lbs/gal
(0.001)
Concentration
mg/L

Mass, lbs (kg)

Mass
lbs (kg)

Volume
MG (m³) | 8.34 lbs/gal
(0.001)
Concentration
mg/L

Volume of Cone

Volume
Cone

1/3 | 0.785
Diameter² | Height

Volume of Cylinder

Volume
Cylinder

0.785 | Height
Diameter²

Volume of Rectangular Tank

Volume
Rectangle

Length | Height
Width

Conversion Factors

1 ac = 4046.9 m^2 or 43 560 sq ft

1 ac ft of water = 326 000 gal

1 atm = 33.9 ft of water

     = 10.3 m of water

     = 14.7 psi

     = 101.3 kPa

1 cfs = 0.646 mgd

     = 448.8 gpm

1 cu ft of water = 7.48 gal

          = 62.4 lb

1 ft = 0.305 m

1 ft H$_2$O = 0.433 psi

1 gal (US) = 3.79 L

       = 8.34 lb of water

1 gr/gal (US) = 17.1 mg/L

1 ha = 10 000 m^2

1 hp = 0.746 kW

     = 746 W

     = 33 000 ft lb/min

1 in. = 25.4 mm or 2.54 cm

1 L/s = 0.0864 ML/d

1 lb = 0.454 kg

1 m of water = 9.8 kPa

1 m^2 = 1.19 sq yd

1 m^3 = 1000 kg

     = 1000 L

     = 264 gal

1 metric ton = 2205 lb

1 mile = 5280 ft

1 mgd = 694 gpm

      = 1.55 cfs

      = 3.785 ML/d

Population equivalent (PE), hydraulic = 378.5 L/cap·d

                            = 100 gpd/cap

PE, organic = 0.077 kg BOD/cap·d

       = 0.17 lb BOD/cap/d

1 psi = 2.31 ft of water

     = 6.89 kPa

1 ton = 2000 lb

1% = 10 000 mg/L

π or pi = 3.14

CHAPTER 1

Introduction to Wastewater Treatment

Components of Wastewater

1. Natural systems have enough assimilative capacity to treat wastewater from urban areas.
   - ☐ True
   - ☐ False

2. Wastewater treatment is necessary to protect public health and the environment.
   - ☐ True
   - ☐ False

3. The term *water resource recovery facility* (WRRF) was recently adopted by Water Environment Federation because it better reflects our goals as a profession.
   - ☐ True
   - ☐ False

4. Most of the fats, oils, and grease present in influent wastewater come from petroleum products.
   - ☐ True
   - ☐ False

5. A pathogen is
   - a. A bacteria or virus found in wastewater
   - b. Any organism capable of causing disease
   - c. Unable to survive for long periods outside wastewater
   - d. Dependent on TSS to reproduce

6. Solids that are retained by a 1.2-μm filter paper and are burned away in a 550 °C furnace are:
   - a. Total dissolved solids (TDS)
   - b. Total volatile solids (TVS)
   - c. Total volatile suspended solids (TVSS)
   - d. Total non-volatile dissolved solids (TVDS)

7. The biochemical oxygen demand (BOD) test measures this:
   - a. Biodegradable organic material
   - b. Percentage of organic suspended solids
   - c. Quantity of live bacteria
   - d. Amount of oxygen needed to stabilize wastewater

8. Solids that are able to pass through a 1.2-μm filter paper and remain unchanged after spending time in a furnace at 550 °C may be described as
   - a. Dissolved and inorganic
   - b. Suspended and inorganic
   - c. Dissolved and organic
   - d. Suspended and organic

9. Some treatment facilities are required to remove ammonia as part of their discharge permits for this reason.
   - a. There is a safe drinking water limit for ammonia.
   - b. Ammonia reacts with organic matter in natural systems to form mustard gas.
   - c. To protect downstream agricultural users from over-fertilizing crops.
   - d. Ammonia is toxic to fish and aquatic life.

10. The biochemical oxygen demand (BOD) test is typically performed as a 5-day test for this reason.
    a. It only takes the Thames River 5 days to reach the ocean.
    b. The bottles only hold enough dissolved oxygen for a 5-day test.
    c. The bacteria only live for 5 days.
    d. All of the organic material is consumed within 5 days.

11. By definition, how much oxygen is required to stabilize or treat 1 kg (lb) of BOD?
    a. 1 kg (lb)
    b. 2 kg (lb)
    c. 3 kg (lb)
    d. 4 kg (lb)

12. Which of the following pollutants is most likely to cause an algae bloom in a lake or river?
    a. Total suspended solids (TSS)
    b. Biochemical oxygen demand (BOD)
    c. Phosphorus
    d. Turbidity

The Clean Water Act

1. The Clean Water Act was promulgated in 1972.
   ☐ True
   ☐ False

2. All direct dischargers are required to have a discharge permit.
   ☐ True
   ☐ False

3. The discharge permit system depends on accurate self-reporting of effluent quality.
   ☐ True
   ☐ False

4. An operator will be fined and serve time in prison if the average effluent suspended solids concentration exceeds the limit given in the discharge permit.
   ☐ True
   ☐ False

5. The monthly discharge monitoring report must be signed by the town attorney.
   ☐ True
   ☐ False

6. U.S. EPA can set limits for heavy metals, cyanide, volatile organic compounds, and pesticides in direct discharge permits.
   ☐ True
   ☐ False

7. A WRRF has a 30-day monthly average BOD_5 limit in their permit of 30 mg/L. Two samples were collected in May with results of 28 and 36 mg/L. The operator should:
   a. Report only the first result to remain below the permit limit.
   b. Average the results together and report a permit violation.
   c. Alter the second result to read 26 mg/L and then average the results together.
   d. Go back to his office and work on his resume.

8. The secondary treatment standards set effluent limits for these parameters:
   a. BOD_5, $CBOD_5$, TSS, and pH
   b. FOG, BOD_5, and TSS
   c. Nitrogen and phosphorus
   d. Pathogenic organisms

9. ABC Corporation manufactures tires in Metro City. All of the process water they generate is discharged to the sewer and is conveyed to the city's water resource recovery facility (WRRF). What type of discharger is ABC Corporation and who issues their discharge permit?
   a. Direct, U.S. EPA
   b. Indirect, state
   c. Indirect, U.S. EPA
   d. Indirect, city WRRF

Liquid Treatment Processes—Preliminary and Primary Treatment

1. Another name for a grinder is a comminutor.
   ☐ True
   ☐ False

2. A trash rack is a manual bar screen with openings between the bars smaller than 5 mm (0.25 in.).
   ☐ True
   ☐ False

3. Which of the following treatment processes would be considered a physical treatment process?
   a. Grit basin
   b. Trickling filter
   c. Chlorine disinfection
   d. Anaerobic digestion

4. Sanitary sewers receive this type of flow.
   a. Stormwater
   b. Municipal wastewater
   c. Both stormwater and municipal wastewater
   d. Both municipal wastewater and industrial wastewater

5. Most municipalities have stopped constructing combined sewers and are removing existing combined sewers for this reason.
   a. Combined sewers deposit raw wastewater in rivers and lakes.
   b. Combined sewers affect WRRF operation during and after storm events.
   c. Combined sewers are difficult to keep clean and can generate odors.
   d. Combined sewers are expensive to construct due to larger pipe diameters.

6. Which technology are you likely to find in a WRRF headworks?
   a. Flow measurement
   b. Primary clarifier
   c. Ultraviolet disinfection
   d. Pond

7. The velocity of wastewater through a rectangular grit basin should be approximately _____ to allow grit to settle while keeping lighter particles in suspension.
   a. 0.15 m/s (0.5 ft/sec)
   b. 0.3 m/s (1.0 ft/sec)
   c. 0.6 m/s (2.0 ft/sec)
   d. 1.5 m/s (5.0 ft/sec)

8. A primary clarifier is capable of removing:
   a. Soluble BOD_5
   b. Ammonia
   c. Total suspended solids
   d. Colloidal solids

9. A WRRF using ponds for secondary treatment is equipped with a manual bar screen that has openings 5-cm (2-in.) apart. What type of debris is most likely to be captured by this screen?

    a. Branches

    b. Rags

    c. Small rocks

    d. Paper

Liquid Treatment Processes—Secondary Treatment

1. Secondary treatment uses chemicals or bacteria to increase the size of particles in wastewater.

    ☐ True

    ☐ False

2. Water resource recovery facilities with secondary treatment typically remove more than 85% of influent BOD_5 and TSS.

    ☐ True

    ☐ False

3. Microorganisms in wastewater consume organic material in the wastewater to sustain themselves and reproduce.

    ☐ True

    ☐ False

4. Secondary treatment systems do not typically include a clarifier or other solids separation processes after biological treatment.

    ☐ True

    ☐ False

5. Ponds are typically unlined.

    ☐ True

    ☐ False

6. A well-functioning activated sludge process will be light tan to dark brown and have a small amount of white to tan foam on the surface.

    ☐ True

    ☐ False

7. A rock media trickling filter should be completely submerged with no free space between the rocks.

    ☐ True

    ☐ False

8. Biological treatment systems use the same microorganisms as natural systems, but are engineered to decrease treatment time by increasing the numbers of microorganisms in the treatment process.

    ☐ True

    ☐ False

9. Disinfection uses chemicals or UV light to sterilize treated wastewater.

    ☐ True

    ☐ False

10. Which of the following treatment processes would be considered biological treatment?

    a. Alum addition for phosphorus removal

    b. Activated sludge

    c. Belt filter press

    d. Ultraviolet disinfection

11. A pond system is categorized as this type of treatment:

    a. Primary

    b. Suspended growth

    c. Fixed growth

    d. Physical

12. Where do the microorganisms in ponds, trickling filters, and activated sludge systems come from?
    a. They are added by the operator.
    b. They are naturally present in the influent wastewater.
    c. They spontaneously generate from suspended solids.

13. This term is used to describe a collection of microorganisms growing on and attached to a media surface such as a rock.
    a. Floc
    b. Slime
    c. Biofilm
    d. Algae

14. In a pond treatment system, what is the purpose of the last pond in the series?
    a. Increases the risk of short-circuiting
    b. Removes the biological solids produced in the first two ponds
    c. Warms the wastewater before discharge
    d. Acts as a primary clarifier or grit basin

15. What is the primary difference between a pond treatment system and an activated sludge system?
    a. Activated sludge recycles settled solids to the beginning of the process.
    b. Pond treatment systems use specialized, cold-tolerant bacteria.
    c. Activated sludge systems use algae for treatment.
    d. Pond treatment systems perform better at higher elevation.

16. For an activated sludge system, which of the following statements is FALSE?
    a. Activated sludge requires less time to treat wastewater than ponds.
    b. Activated sludge is a suspended growth biological process.
    c. Activated sludge uses fungus to treat wastewater.
    d. Activated sludge holds the biological solids longer than the wastewater.

17. An example of a fixed-film treatment process is
    a. Activated sludge
    b. Pond
    c. Rotating biological contactor
    d. Clarifier

18. Which two methods of disinfection are most commonly used in domestic WRRFs?
    a. Ozone and chlorine
    b. Chlorine and UV light
    c. Bleach and ozone
    d. Ultraviolet light and boiling

19. Match the unit process to its place in the liquid treatment side.

    1. Collection system   a. Pretreatment
    2. Grit basin   b. Primary treatment
    3. Primary clarifier   c. Disinfection
    4. Activated sludge   d. Preliminary treatment
    5. Chlorine addition   e. Secondary treatment

20. Draw a line from the liquid treatment type to its treatment goal.

    1. Preliminary treatment   a. Reduce number of bacteria and pathogens
    2. Primary treatment   b. Increase particle size for separation step
    3. Secondary treatment   c. Protect downstream equipment
    4. Tertiary treatment   d. Remove nitrogen and phosphorus
    5. Disinfection   e. Decrease size and cost of secondary treatment

Solids Treatment Processes

1. Sludge stabilization processes like digesters transform sludge into biosolids.
   - ☐ True
   - ☐ False

2. Primary sludge typically contains between 2 and 6% total solids.
   - ☐ True
   - ☐ False

3. The beneficial use of biosolids is regulated under 40 CFR part 258.
   - ☐ True
   - ☐ False

4. The time required for aerobic digestion depends on the temperature of the sludge.
   - ☐ True
   - ☐ False

5. Put the following solids treatment processes into the correct order.
   a. Dewatering
   b. Secondary clarifier
   c. Thickening
   d. Digestion

6. Screenings and grit are typically:
   a. Sent to a landfill
   b. Dewatered and land applied
   c. Used for road base
   d. Digested anaerobically

7. Primary sludge consists of
   a. Microorganisms grown during treatment
   b. Rags, plastic, and other heavy materials
   c. Unprocessed, settleable organic and inorganic solids
   d. Grit and screenings

8. Secondary sludge consists of
   a. Microorganisms grown during treatment
   b. Rags, plastic, and other heavy materials
   c. Unprocessed, settleable organic and inorganic solids
   d. Grit and screenings

9. This type of biosolids may be made available for public takeaway.
   a. Class A
   b. Class B
   c. Class C
   d. Class D

10. The vector attraction reduction requirement in the biosolids 503 regulations
    a. Limits concentrations of heavy metals in biosolids
    b. Allows screenings and grit to be comingled with digested sludge
    c. Reduces the likelihood that rats and insects will be attracted to finished biosolids
    d. Prevents application of biosolids near streams and lakes

11. Sludge thickening and biosolids dewatering are performed for this reason:
    a. Reduces the total volume of sludge or biosolids
    b. Required by the discharge permit
    c. Reduces the total mass of sludge or biosolids
    d. Required by the 503 regulations

12. All of the following statements about anaerobic digestion are true EXCEPT:
    a. Reduces the amount of biosolids
    b. Meets vector attraction reduction requirements
    c. Typically paired with primary clarifiers
    d. Break down solids in the presence of oxygen

13. Anaerobic digester gas contains approximately
    a. 70% nitrogen and 22% oxygen
    b. 65% methane and 35% carbon dioxide
    c. 70% nitrogen and 30% carbon dioxide
    d. 65% methane and 22% oxygen

14. Match the process to its defining characteristic.
    1. Aerobic digestion   a. Uses lime to increase sludge pH
    2. Anaerobic digestion   b. Takes place prior to digestion
    3. Thickening   c. Breaks down sludge with oxygen
    4. Dewatering   d. Produces methane gas
    5. Chemical stabilization   e. Produces "cake" of up to 50% solids

15. This piece of equipment rapidly spins biosolids to remove water.
    a. Belt filter press
    b. Rotary drum thickener
    c. Centrifuge
    d. Heat dryer

CHAPTER 2
Wastewater Characteristics

Sources of Wastewater

1. The Federal Pretreatment Act defines industrial and commercial users differently from one another.
    - ☐ True
    - ☐ False

2. Influent wastewater consists of domestic, industrial, and commercial wastewater as well as contributions from inflow and infiltration.
    - ☐ True
    - ☐ False

3. Influent flow patterns reflect activities taking place in the service area.
    - ☐ True
    - ☐ False

4. An operator must take a piece of equipment out of service for maintenance. They will need to bypass pump wastewater around this piece of equipment during repairs. Assuming the WRRF has a typical diurnal flow pattern for domestic wastewater, when should maintenance be scheduled to minimize the amount of bypass pumping?
    - a. Midmorning
    - b. After lunch
    - c. Afternoon
    - d. Late evening

5. Which type of service area is likely to see the greatest variations in influent flow over a single day?
    - a. Town with 500 residents
    - b. City with separate domestic and storm sewers
    - c. City with more than 50,000 residents
    - d. Town without large commercial or industrial users

6. A WRRF receives wastewater from two ski resorts. Assuming the number of full-time residents in the service area is constant year-round, when should the operator expect to see the highest influent flows and loads?
    - a. Spring
    - b. Summer
    - c. Fall
    - d. Winter

Influent Characteristics

1. When water use increases in a community, the influent wastewater concentrations decrease assuming that the population served has not changed.
    - ☐ True
    - ☐ False

2. An influx of stormwater from a combined sewer can increase average influent wastewater temperatures.
    - ☐ True
    - ☐ False

3. Septic conditions in the collection system upstream of the influent flow to a WRRF can cause the influent to be darker than usual and smell like rotten eggs.
   ☐ True
   ☐ False

4. Conductivity meters can also be used to accurately measure total dissolved solids.
   ☐ True
   ☐ False

5. Domestic wastewater includes contributions from
   a. Significant industrial users
   b. Categorical industrial users
   c. Homes and businesses
   d. Heavy industrial users

6. Fresh domestic wastewater will typically be
   a. White to pale yellow
   b. Light brown to grey
   c. Gray to black
   d. Orange, yellow, or blue

7. Hydrogen sulfide is a concern for all of these reasons EXCEPT
   a. Poisonous at low concentrations
   b. Corrodes concrete and metal
   c. Potentially explosive
   d. Smells strongly of garlic

8. Turbidity is a measurement of:
   a. Light scatter
   b. Cloudiness
   c. Solids concentration
   d. Organic matter

9. Alkalinity is a measurement of
   a. pH
   b. Buffering capacity
   c. Calcium carbonate concentration
   d. Hydroxide content

10. Which of the following pH values would be considered acidic?
    a. 4.6
    b. 7.1
    c. 8.3
    d. 9.4

11. If all of the alkalinity is consumed, what will the pH be?
    a. 1.2
    b. 4.5
    c. 7.0
    d. 8.3

12. An influent sample is analyzed for both COD and BOD. Which of the following statements must be true?
    a. BOD is equal to or greater than COD
    b. The BOD test was completed before the COD test
    c. COD is equal to or greater than BOD
    d. The COD test was performed at 20 °C

13. The laboratory reported a phosphorus concentration in the final effluent of 2.5 mg/L as PO_4^{-3}. What is this in milligrams per liter of PO_4-P?
    a. 0.25 mg/L PO_4-P
    b. 0.81 mg/L PO_4-P
    c. 2.5 mg/L PO_4-P
    d. 7.7 mg/L PO_4-P

CHAPTER 3
Preliminary Treatment of Wastewater

Introduction to Preliminary Treatment

1. All WRRF headworks include screening, grit removal, and flow measurement.
    - ☐ True
    - ☐ False

2. Screening and grit removal must be completed before flow measurement.
    - ☐ True
    - ☐ False

3. This device may be used to direct flow to or around a screen or grit basin.
    a. Cogwheel
    b. Slide gate
    c. Weir block
    d. Fraser valve

4. A bar screen may not prevent this type of material from entering the WRRF.
    a. Sticks
    b. Large rocks
    c. Rags
    d. Pallets

5. Grit basins typically remove sand, gravel, eggshells, and coffee grounds by
    a. Placing wire mesh in the flow path as a strainer
    b. Scooping the surface of the water
    c. Introducing microorganisms to consume them
    d. Decreasing water velocity and allowing them to settle

Screening Purpose and Design

1. Removal of grit, rags, and other debris is often necessary to protect downstream equipment from harm.
    - ☐ True
    - ☐ False

2. If too much organic material is removed at the WRRF headworks, the downstream biological process can be affected.
    - ☐ True
    - ☐ False

3. Head loss is the difference in water level between the upstream and downstream sides of a screen.
    - ☐ True
    - ☐ False

4. One consequence of allowing excess organic material to be removed along with rags and other inert debris is
    a. Oxygen generation
    b. Reduced capture of rags
    c. Generation of odors
    d. Reduced disposal costs

5. This type of screen has the widest openings and only stops the largest debris.
   a. Trash rack
   b. Manual bar screen
   c. Mechanical bar screen
   d. Step screen

6. A WRRF currently has a bar screen with 50-mm (2-in.) openings. Operators of the WRRF are considering replacing the screen with one that has 25-mm (1-in.) openings. How much should they expect the volume of screenings removed to change?
   a. Volume will remain about the same
   b. Screening volume will double
   c. Screening volume will increase by a factor of 4
   d. Screening volume will decrease by 50%

7. What is the most likely effect of receiving large quantities of grease on a perforated plate-type screen?
   a. Blinding of the screen
   b. Decreased cleaning frequency
   c. Passage of hair and rags into the WRRF
   d. Improved cog lubrication

8. Place the following screen types in order from the smallest opening size to the largest.
   a. Fine screen
   b. Trash rack
   c. Bar screen
   d. Micro screen

Screening Equipment

1. Trash and bar racks are installed at a 30- to 45-deg angle from the vertical.
   ☐ True
   ☐ False

2. Mechanical cleaning of bar screens increases labor costs because of increased maintenance.
   ☐ True
   ☐ False

3. Fine screens capture fewer organic solids than coarse screens.
   ☐ True
   ☐ False

4. Rollback occurs when screened material falls back down the face of the screen, possibly being returned to the wastewater flow.
   ☐ True
   ☐ False

5. A pond WRRF is most likely to have this type of screen in its headworks.
   a. Manual bar screen
   b. Perforated plate screen
   c. Stair screen
   d. Micro screen

6. Match the screen type to its identifying characteristic.

   1. Trash rack
   2. Manual bar screen
   3. Single rake screen
   4. Catenary screen
   5. Continuous self-cleaning screen
   6. Perforated plate screen
   7. Stair screen
   8. Inclined cylindrical screen

   a. Single operating cycle
   b. Interlocking links, no lower sprocket
   c. Efficiently removes hair and stringy material
   d. Integrate wash press
   e. Remove only largest debris
   f. Rectangular openings limited in height and width
   g. May be located in a bypass channel
   h. Two sets of lamella lift solids up stairs

7. On a single rake screen, a limit switch
   a. Sets the upper and lower limits of travel for the rake
   b. Adjusts input line voltage
   c. Modifies the minimum particle size captured
   d. Turns off the drive motor if the rake is jammed

8. One disadvantage of using comminutors and grinders is
   a. Reduced potential for clogged pipes and damaged equipment
   b. Increased screenings disposal costs
   c. Shredded material reduces treatment capacity downstream
   d. More frequent overflows of the influent channel

Screening Process Control, Operation, and Maintenance

1. A WRRF with a combined sewer system is expecting a large storm. The manual bar screens are normally cleaned every hour. The operator should plan to decrease the cleaning frequency during and immediately after the storm.
   ☐ True
   ☐ False

2. A shear pin is designed to break under a certain amount of force or pressure to prevent additional damage to equipment.
   ☐ True
   ☐ False

3. Screens should be cleaned before the head loss across the screen reaches _____ or according to the screen manufacturer's recommendations.
   a. 2.5 cm (1 in.)
   b. 7.6 cm (3 in.)
   c. 12.7 cm (5 in.)
   d. 17.8 cm (7 in.)

4. This type of control mechanism is best suited for storm events or other conditions that produce highly variable amounts of screenable material.
   a. Clock timer
   b. Level sensor
   c. Particle counter
   d. HOA switch

5. A mechanical bar screen must be taken out of service for maintenance. Put the following steps in the correct order.
   a. Shut down the screen according to the manufacturer's directions.
   b. Close the upstream (inlet) gate.
   c. Close the downstream (outlet) gate.
   d. Wash the screen to remove debris.
   e. Drain the channel and remove debris.
   f. Lockout and tagout equipment.
   g. Wash sides of channel and perform final draining of channel.

6. At a minimum, how often should screens be inspected for visible and audible indications of possible malfunctions?
   a. Daily
   b. Weekly
   c. Monthly
   d. Quarterly

Grit Removal Purpose and Design

1. The volume of grit removed for every m³ (mgd) of flow received is very consistent from one facility to the next.
   ☐ True
   ☐ False

2. As flow velocity through the basin increases, larger, denser particles will pass through to the next process.
   ☐ True
   ☐ False

3. All of the following characteristics will influence the settling velocity of a particle EXCEPT
   a. Density
   b. Size
   c. Color
   d. Shape

4. Gravity pulls grit particles straight down, but this pushes them forward toward the outlet of the grit basin:
   a. Velocity
   b. Density
   c. Viscosity
   d. Conductivity

5. Which of the following particles should settle the fastest through a column of water?
   a. Corn kernel
   b. Coffee grounds
   c. Sand
   d. Gravel

6. A grit basin has a flow velocity of 0.3 m/s (1 ft/sec) at 9 A.M. As flows increase throughout the day, grit basin performance will be affected in this way.
   a. The percentage of organic material captured will increase.
   b. Grit removal efficiency will decrease.
   c. Velocity through the basin will decrease.
   d. Air consumption will increase.

Grit Removal Equipment

1. Velocity is calculated by multiplying the flowrate by the cross-sectional area.
   ☐ True
   ☐ False

2. Grit classifiers return organic material to the wastewater flow and remove excess water from the grit.
   ☐ True
   ☐ False

3. A WRRF currently has one grit basin in service. The flow velocity through the basin is 0.8 m/s (2.6 ft/sec). How many grit basins should be in service to maintain a flow velocity between 0.24 and 0.3 m/s (0.8 and 1.0 ft/sec)?
   a. 1
   b. 2
   c. 3
   d. 4

4. Flow-proportional weirs are used in velocity control horizontal flow grit basins for this purpose:
   a. Produce a constant discharge rate
   b. Reduce buildup of grit behind the weir
   c. Measure flowrate
   d. Maintain a constant velocity

5. In an aerated grit basin, this process control parameter influences which particles will sink and which will continue on to the next process.
   a. Velocity of the roll pattern
   b. Depth of the basin
   c. Vortex dimensions
   d. Flow-proportional weir setting

6. This type of grit basin may use paddles to maintain the flow circulation pattern.
   a. Detritus tank
   b. Aerated grit basin
   c. Vortex grit basin
   d. Velocity control

7. An operator must bypass their aerated grit basin for several days while annual maintenance is performed on the diffusers. Which of the following statements is true?
   a. Airflow should be set to 50% of maximum after basin dewatering.
   b. Grit may pass through and cause damage to downstream equipment.
   c. The organic loading to downstream processes will increase.
   d. Screenings will accumulate in the grit channel.

8. Grit is typically removed from a vortex-type grit basin in this way.
   a. Manual removal with shovels
   b. Chain-and-flight system
   c. Screw auger
   d. Grit pumps

Grit Removal Process Control, Operation, and Maintenance

1. Increased equipment run time will increase O&M costs.
   ☐ True
   ☐ False

2. When grit pumps are used to remove grit from a hopper, a timed cycle of 10 minutes of operation once every 30 minutes should always be used.
   ☐ True
   ☐ False

3. When placing a rectangular grit basin with a chain-and-flight system into service, the basin should be filled completely before starting the mechanism.
   ☐ True
   ☐ False

4. A sample of grit is grayish in color and smells like rotten eggs. The operator should consider
    a. Slowing down the vortex paddle
    b. Increasing flow velocity through the grit basin
    c. Decreasing the rinse water to the grit classifier
    d. Placing another grit basin into service

5. The lower port of a grit cyclone is partially blocked by a piece of trash. The blockage will likely cause
    a. Excessive vibration of the cyclone
    b. Increasing flow velocity through the grit basin
    c. Decreasing the rinse water to the grit classifier
    d. Liner delamination

6. The headworks building at a new WRRF shows evidence of metal and concrete corrosion. The problem seems to be even worse in covered channels. The most likely cause is
    a. Industrial discharges
    b. Grease buildup
    c. Inadequate ventilation
    d. Over-aeration

7. An operator is used to seeing an even mixing pattern on the surface of the aerated grit basin. Today, the surface is fairly smooth and does not show signs of good mixing. The operator should
    a. Increase the airflow rate to the basin
    b. Check the diffusers for rags
    c. Replace broken headers and diffusers
    d. Adjust the outlet weir

Additional Pretreatment Considerations

1. The water level in flow equalization basins should fluctuate over a 24-hour period.
    ☐ True
    ☐ False

2. Because septic waste is already partially treated, it tends to be lower strength than typical domestic wastewater.
    ☐ True
    ☐ False

3. Septage can contain rocks and other debris, which can damage downstream equipment if it is not screened out.
    ☐ True
    ☐ False

4. Grease trap waste can be accepted and treated by most WRRFs under the 503 Regulations as long as it comes from restaurants and not categorical industrial users.
    ☐ True
    ☐ False

5. A septage hauling and receipt manifest should include all of the following information EXCEPT
    a. Source, type, and amount of septage
    b. Signed certification by the generator
    c. Hauler's commercial driver's license number
    d. Septage receipt date

6. Septage wastes received by WRRFs should be monitored and tracked to
    a. Proportion cost of service to customers
    b. Ensure treatment capacity is not exceeded
    c. Calculate percentage of industrial septage received
    d. Meet paint filter test requirements

7. This test measures the amount of free water in screenings and grit:
    a. Total suspended solids
    b. Mesh size
    c. Biochemical oxygen demand
    d. Paint filter

CHAPTER 4
Primary Treatment of Wastewater

Introduction to Primary Treatment of Wastewater

1. Primary clarifiers are capable of removing 100% of the influent TSS and BOD_5.
   - ☐ True
   - ☐ False

2. Which of the following components is most likely to be in the sludge at the bottom of a primary clarifier?
   a. Ammonia
   b. Soluble BOD
   c. Grease
   d. Settleable solids

3. The main purpose of primary treatment is to
   a. Remove grit and larger debris to protect downstream equipment.
   b. Reduce the loading to the secondary treatment process.
   c. Convert particulate BOD into soluble BOD for easier treatment.
   d. Generate electricity by reducing the organic load to the digester

Theory of Operation

1. Clarifiers rely on density differences to separate particles from wastewater.
   - ☐ True
   - ☐ False

2. Two particles have the same density. The particle with more surface area will settle faster than the particle with less surface area.
   - ☐ True
   - ☐ False

3. Primary sludge will continue to compact and achieve higher total solids concentrations for as long as it remains in the clarifier.
   - ☐ True
   - ☐ False

4. Which of the following particles should reach the bottom of the clarifier first?
   a. Grease
   b. Feather
   c. Sand
   d. Lettuce

5. What phenomena in a clarifier push particles up and out of the clarifier?
   a. Drag
   b. Surface tension
   c. Gravity
   d. Surface overflow rate

6. Surface overflow rate is comparable to
   a. Detention time
   b. Velocity
   c. Weir loading rate
   d. Density

7. The sludge layer that forms at the bottom of the clarifier is called the
   a. Schmutzdecke
   b. Float
   c. Blanket
   d. Emulsion

8. A potential negative consequence of leaving primary sludge in the clarifier for an extended period of time might be
   a. Hydrogen sulfide gas
   b. Higher solids concentrations
   c. Lower solids concentrations
   d. Breakdown of grit particles

Design Parameters

1. A well-operated primary clarifier will always remove 40% of total BOD_5.
   ☐ True
   ☐ False

2. Primary clarifier performance is limited by SOR, HDT, and the characteristics of the incoming wastewater.
   ☐ True
   ☐ False

3. Biodegradable organic solids are measured by both the BOD and VSS laboratory tests.
   ☐ True
   ☐ False

4. The influent to a primary clarifier contains 190 mg/L of TSS. The effluent contains 125 mg/L of TSS. Find the percent removal.
   a. 26.1%
   b. 34.2%
   c. 65.7%
   d. 87.4%

5. The influent to a primary clarifier contains 250 mg/L of BOD. Fifty milligrams per liter is soluble and 200 mg/L is particulate. If the clarifier removes 50% of the particulate BOD, what will the total BOD concentration be in the primary effluent?
   a. 50 mg/L
   b. 100 mg/L
   c. 125 mg/L
   d. 150 mg/L

6. A typical HDT for a primary clarifier would be
   a. 30 minutes
   b. 1.5 hours
   c. 3 hours
   d. 6 hours

Equipment—Circular Clarifiers

1. Fillets are used to fill in the corners of square and octagonal clarifiers.
   - ☐ True
   - ☐ False

2. Imhoff tanks settle primary sludge in the upper compartment and digest it in the lower compartment.
   - ☐ True
   - ☐ False

3. Generally, 1.5 m (5 ft) of freeboard is provided. This allows excess water to be stored when there is a blockage or downstream equipment malfunction.
   - ☐ True
   - ☐ False

4. This device helps prevent floatable material from escaping over the weir.
   - a. Energy dissipater
   - b. Rake
   - c. Scum baffle
   - d. Wear strip

5. In a circular primary clarifier, the sludge hopper is typically located
   - a. Along the outer edge of the clarifier
   - b. Near the center of the clarifier
   - c. Adjacent to the scum baffle
   - d. Below the scum box

6. Most primary clarifiers are smaller than 100 m (300 ft) in diameter for this reason:
   - a. Wind can create currents in larger clarifiers
   - b. Larger sludge collection mechanisms are not available
   - c. Difficulty in reaching equipment for cleaning
   - d. Skimmer arm weight cannot be supported

7. Most circular, primary clarifiers use a center-feed, peripheral-rim, take-off flow pattern.
   - ☐ True
   - ☐ False

8. The feedwell directs flow down toward the bottom of the clarifier while increasing flow velocity.
   - ☐ True
   - ☐ False

9. A WRRF has three primary clarifiers, but only two of them are currently in service. The influent velocity is currently 2 m/s (6.6 ft/sec). The operator should
   - a. Record the influent velocity on the log sheet.
   - b. Place the third clarifier into service.
   - c. Adjust the feedwell baffles.
   - d. Take a clarifier out of service.

10. The purpose of the feed well in a primary clarifier is to
    - a. Prevent short-circuiting
    - b. Channel the flow to the effluent weir
    - c. Aerate and mix the influent wastewater
    - d. Collect scum and sludge

Label the components of this circular primary clarifier. Check your answers against the filled-out version of this figure on p. 129 of the Answer Key.

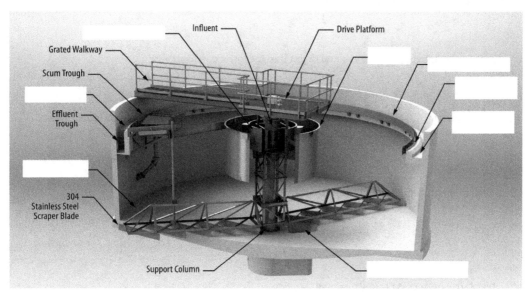

Reprinted with permission by Monroe Environmental (Monroe, MI) www.mon-env.com

11. Weirs for circular clarifiers generally consist of rectangular openings spaced at 0.3-m (1-ft) intervals.
    ☐ True
    ☐ False

12. The plows on a circular clarifier scraper mechanism are angled to direct sludge to the center hopper.
    ☐ True
    ☐ False

13. The drive for the sludge collection mechanism is typically mounted on the wall of a circular clarifier.
    ☐ True
    ☐ False

14. This type of launder projects into the clarifier center and is mounted on the outer wall.
    a. Inset launder
    b. Outboard launder
    c. Cantilevered launder
    d. Inboard launder

15. One potential disadvantage of an inboard launder is
    a. Weir length is increased compared to an inset launder
    b. Lack of a mounting location for the weir plate
    c. Floatable material can become trapped under the launder
    d. Difficulty in reaching both weir plates for cleaning

16. One advantage of using an inset or cantilevered launder is
    a. Increases total weir length
    b. Difficulty reaching weir plates for cleaning
    c. Floatable material can become trapped
    d. Increases weir loading rate

17. Scum and other floatable material
    a. Passes under the scum baffle and into the effluent.
    b. Is pushed by the skimmer arm onto the egress ramp.
    c. Consists primarily of grease and sand.
    d. Should be removed upstream in preliminary treatment.

18. Clarifier drive units typically consist of a motor and three sets of reducers.
    ☐ True
    ☐ False

19. A sprocket is a toothed wheel that is used in combination with a chain or belt.
    ☐ True
    ☐ False

20. Gears transfer power from one machine to another.
    ☐ True
    ☐ False

21. A clarifier drive mechanism has a broken shear pin. The operator should
    a. Replace the shear pin and restart the drive.
    b. Determine and correct the source of the overload condition.
    c. Decrease the drive motor speed.
    d. Adjust the gear-to-pinion teeth ratio.

22. The clarifier mechanism is hung up on the clarifier floor. The resulting torque load reaches 45% of the maximum design drive torque. As a result, the worm gear shifts inside the gearbox causing
    a. Destruction of the spur gear
    b. Drive motor to shut down
    c. Rake speed to decrease
    d. Overtorque alarm to trigger

23. When two gears are placed next to one another, the smaller gear is called the
    a. Gear
    b. Idler
    c. Pinion
    d. Worm

24. This type of gear may be used when it is necessary to connect a horizontal shaft to a perpendicular gear.
    a. Sprocket
    b. Worm
    c. Idler
    d. Pinion

25. Two gears are operating together. The smaller gear is spinning at 2000 rpm. The larger gear spins at 500 rpm. How much more torque does the larger gear have?
    a. Both gears have the same torque
    b. Twice as much
    c. Four times less
    d. Eight times as much

26. When a reducer is placed between a motor and a piece of equipment
    a. Speed is reduced and torque is increased.
    b. Speed is increased and torque is reduced.
    c. Speed is reduced and torque is reduced.
    d. Speed is increased and torque is increased.

Equipment—Rectangular Clarifiers

Label the following components of a rectangular clarifier. Check your answers against the filled-out figure on p. 130 of the Answer Key.

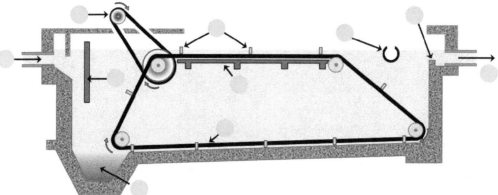

1. The flow in a rectangular clarifier is radial.
    ☐ True
    ☐ False

2. A chain-and-fight system consists of two, continuous, parallel loops of chains.
    ☐ True
    ☐ False

3. A perforated inlet baffle may be used to break up jetting action and disperse flow in a rectangular clarifier.
    ☐ True
    ☐ False

4. This type of sludge collection mechanism is commonly used with rectangular clarifiers.
    a. Niersfratzer spiral
    b. Chain and flight
    c. Window shade plow
    d. Logarithmic spiral plow

5. When a high-high torque alarm is triggered, this may also occur.
    a. Tooth loss on the pinion gear
    b. Collector mechanism may speed up
    c. Load indicator shows zero
    d. Shear pin breaks

6. A minimum of _____ sprockets is needed to support a chain-and-flight system that removes both settleable and floatable material.
    a. 1
    b. 2
    c. 3
    d. 4

7. The drive unit for a rectangular clarifier is located
    a. On the clarifier wall
    b. Underneath the inlet baffle
    c. Parallel to the scum skimmer
    d. In the influent manifold

8. The inlet manifold
   a. Is typically 25% of the tank width
   b. Distributes flow evenly across the tank
   c. Requires a minimum of 10 openings
   d. Incorporates a waterfall cascade

9. One disadvantage of submerged launders over transverse weirs is
   a. Diminished algae growth
   b. Easier to remove scum from the clarifier surface
   c. Trapped rags and debris can be difficult to remove
   d. Higher flow velocities

10. Plastic or metal wear shoes at either end of a clarifier flight
    a. Support the weight of the flights
    b. Do not require replacement
    c. Compensate for stretching of the chain
    d. Assist in cleaning the scum dipper

11. Historically, cast iron chains and wooden flights were used. Today, most systems use stainless steel or plastic with fiberglass flights for this reason.
    a. Plastic chains are less likely to decay in sunlight.
    b. Cast iron is more expensive.
    c. U.S. Environmental Protection Agency requires the use of food-grade materials.
    d. These materials are lighter and easier to maintain.

12. A transverse collection trench may be used in this situation.
    a. Longitudinal trench will not fit the space
    b. Length-to-width ratio is less than 3:1
    c. Multiple chain-and flight systems used in parallel
    d. Expected sludge concentrations exceed 10%

13. The trough rotation on a rotating skimmer
    a. Typically operates on a timed cycle
    b. Turns in the direction of flow towards the weir
    c. May be triggered by an approaching skimmer arm
    d. Cannot be used with submerged launders

Process Variables

1. Hydraulic detention time can be defined as the amount of time, on average, water remains in a vessel.
   ☐ True
   ☐ False

2. As SOR increases, the velocity of water through the clarifier decreases.
   ☐ True
   ☐ False

3. A primary clarifier has a diameter of 36.6 m (120 ft) and a depth of 4.57 m (15 ft). Find the volume of the clarifier. _____

4. The influent flow to a primary clarifier is 30.3 ML/d (8 mgd). If the clarifier holds 2.27 ML (0.6 mil. gal), what is the HDT in hours? _____

5. Find the SOR for a primary clarifier with a diameter of 36.6 m (120 ft) and an influent flow of 30.3 ML/d (8 mgd). _____

6. All of the following variables can affect the settling velocity of a particle EXCEPT
   a. Size
   b. Shape
   c. Density
   d. Color

7. A long, warm collection system may have this effect on the influent wastewater:
   a. Decreases particle size
   b. Increases ratio of inorganic to organic particles
   c. Decreases ammonia concentration
   d. Increases opportunities for flocculation

8. As water temperatures decrease in a primary clarifier,
   a. Particles will settle slower
   b. Odor potential increases
   c. Particles will settle faster
   d. Water becomes less viscous

9. A small WRRF decides to accept hauled septage waste from a nearby campground. On days when septage is received
   a. Percent removal of TSS increases
   b. Percent removal of soluble BOD increases
   c. Percent removal of both TSS and BOD decreases
   d. Percent removal is unaffected

10. Cold water affects settling velocity because
    a. Cold water contains more dissolved gases
    b. There is less drag on particles
    c. Microbial activity increases as temperatures drop
    d. Water is denser at colder temperatures

11. These conditions tend to entrain grease and prevent it from being skimmed off in the primary clarifier.
    a. Presence of ducks
    b. Low pH, high temperatures
    c. Lack of turbulence in the sewer
    d. High pH, low temperatures

Process Control—Hydraulic Considerations and Solids Handling

1. Sludge blanket depths should generally be kept below 0.9 m (3 ft).
   ☐ True
   ☐ False

2. Managing the collection system can help reduce odors at the WRRF.
   ☐ True
   ☐ False

3. Given the following information, calculate the mass of dry solids that will end up in the clarifier blanket over a 24-hour period. Primary influent TSS is 280 mg/L and primary effluent TSS is 168 mg/L. The average daily flow through the WRRF is 12.5 ML/d (3.3 mgd). _____

4. A primary clarifier captures 2671 kg (5885 lb) of solids. If the total solids concentration in the sludge is 4.2%, how much volume will need to be pumped? Express your answer as m^3 (gal). _____

5. A primary clarifier produces 20.2 m^3 (5333 gal) of primary sludge at a concentration of 4.2% total solids. If the sludge pump has a capacity of 435 L/min (115 gpm), how many minutes will the sludge pump need to operate each day? _____

6. Thickened sludge concentrations greater than 6% total solids
    a. Are easily achieved in primary clarifiers.
    b. Can cause problems with sludge collection and pumping.
    c. Deliver excess water to anaerobic digesters.
    d. Would be considered "thin" sludge.

7. The primary sludge concentration is approaching 8% total solids. A high-torque alarm has sounded for the drive mechanism. The operator should
    a. Replace the shear pin with a stronger model.
    b. Decrease the clarifier mechanism speed.
    c. Increase the sludge pumping rate to decrease the total solids concentration.
    d. Remove a clarifier from service to decrease detention time.

8. A WRRF begins adding ferric chloride to its influent channel for odor control. Primary sludge production
    a. Increases
    b. Decreases
    c. Is unaffected
    d. Doubles

9. When sludge pumping is intermittent
    a. The total solids concentration in the sludge tends to be lower
    b. Sludge lines should be flushed with water
    c. The risk of septicity and hydrogen sulfide generation is lower
    d. Pump cycle times should be coordinated with sludge hopper size

10. This week, the percentage of BOD removal over the clarifier has dropped from 40 to 25%; however, the percent removal of TSS is unchanged. What is the most likely cause?
    a. Recycling of solids and BOD from the anaerobic digester
    b. Industrial discharge of soluble BOD
    c. Temperature changes affecting settling velocities
    d. Collection mechanism is broken

Process Control—Skimming, Upstream, and Downstream Units

1. Chemical addition for odor control and phosphorus removal upstream of the primary clarifiers can increase the percentage of BOD and TSS removed by the primary clarifier.
    ☐ True
    ☐ False

2. Recycle flows from other WRRF processes are best returned to the primary clarifiers during peak hour influent flow.
    ☐ True
    ☐ False

3. The surface of a primary clarifier has a buildup of scum and foam and smells strongly of rotten eggs. The operator should
    a. Adjust the sludge pumping frequency
    b. Remove the scum baffle
    c. Increase the skimming frequency
    d. Verify proper operation of the sludge pump

4. A poorly functioning grit basin or bar screen
    a. Retains excessive quantities of grit
    b. Reduces odors upstream of the primary clarifier
    c. Generates soluble BOD
    d. May pass debris on to the primary clarifier

5. A WRRF has two primary clarifiers. One must be taken out of service for cleaning and maintenance. While this is being done, the SOR in the remaining clarifier will be excessively high. What is the likely effect to overall facility operations?

    a. Longer HRTs through the remaining clarifier

    b. Increased BOD and TSS loading to the secondary treatment process

    c. Reduced efficiency of the grit basin and screening equipment

    d. Increased BOD and TSS loading to the solids handling process

Operation

1. All rectangular clarifier mechanisms may be safely operated in a completely dry tank.

    ☐ True

    ☐ False

2. When placing a circular clarifier into service, the mechanism should be operated for several revolutions in a dry tank to verify correct operation.

    ☐ True

    ☐ False

3. When a primary clarifier is taken out of service, it is acceptable to leave sludge in pipelines and pumps.

    ☐ True

    ☐ False

4. A rectangular clarifier with a chain-and-flight sludge collection mechanism will be placed into service. Because the plastic chain is several years old, the operator may need to

    a. Lubricate the chain with a lightweight machine oil

    b. Ensure the chain has not rusted to the sprockets

    c. Remove a link from each chain to compensate for wear

    d. Replace the wear shoes on each flight

5. When removing a tank from service, it should be immediately flushed with water for this reason.

    a. Prevent generation of odors

    b. Dried sludge is harder to remove

    c. Continued operation of water lubricated bearings

    d. Tank may be returned to service faster

6. A sample is collected from the primary clarifier effluent for the settleable solids test. After settling for an hour, the tip of the cone is empty. The operator concludes

    a. Excessive grit is entering the clarifier

    b. The sludge collector speed should be increased

    c. Settleable solids are passing on to the secondary process

    d. Process efficiency cannot be improved

7. Routine laboratory testing shows that the percentage of volatile solids in the primary sludge has decreased over the last week from 80% down to 70%. What is the most likely cause?

    a. Increased BOD loading to the WRRF

    b. Failure of upstream grit removal equipment

    c. Sludge blanket is deeper than 1.2 m (4 ft)

    d. Hydraulic retention time should be decreased

Maintenance

1. Clarifiers should be drained and all equipment and concrete thoroughly inspected annually.

    ☐ True

    ☐ False

2. Condensate must be drained from clarifier drive units.
   a. Daily
   b. Weekly
   c. Monthly
   d. Annually

3. Weirs should be kept clear of debris, level, and clean to prevent
   a. Uneven flow distribution
   b. Retention of fine particulates
   c. Unsightly growth of algae
   d. Clogging of the flush valves

4. Each facility should create its own equipment maintenance logs. What is the best source of information for determining the type and frequency of necessary maintenance?
   a. Books like this one that provide general information
   b. Online training classes that provide general information
   c. Manufacturer's operations and maintenance manuals
   d. Chilton's guide

Troubleshooting

1. Primary sludge concentrations are between 1.5 and 3%. What action can the operator take to increase the sludge concentration?
   a. Decrease pumping rate
   b. Increase collector speed
   c. Remove a clarifier from service
   d. Clean the sludge lines

2. The settleable solids test shows a significant amount of settlable material in the primary effluent. What action can the operator take to reduce effluent settleable solids?
   a. Adjust recycle flows to coincide with peak hour flow
   b. Decrease collector speed
   c. Place additional clarifiers into service
   d. Adjust sludge withdrawal rate

3. The shear pin on a chain-and-flight system has broken three times over the previous week. The collector is operated intermittently. To prevent additional breakages, the operator could
   a. Install a stronger shear pin
   b. Increase the collector operating time
   c. Decrease sludge withdrawal rate
   d. Remove a clarifier from service

4. Over the past month, it has become progressively more difficult to remove sludge from the hopper. Loading to the WRRF has not changed. The upstream headworks equipment is functioning well. What is the most likely cause?
   a. Decreased grit content in the sludge
   b. Velocity in sludge lines is too high
   c. Squeegees on flights need to be replaced
   d. Buildup of grease or other material in the sludge lines

CHAPTER 5

Fundamentals of Biological Treatment

Introduction to Biological Treatment

1. Biological secondary treatment processes contain bacteria in much higher concentrations than natural systems.
   - ☐ True
   - ☐ False

2. Individual bacteria are easily observed with the naked eye.
   - ☐ True
   - ☐ False

3. The 5-day CBOD test measures all of the biodegradable material present in the wastewater.
   - ☐ True
   - ☐ False

4. Liquid may be adsorbed into a sponge.
   - ☐ True
   - ☐ False

5. One goal of biological treatment systems is to
   - a. Increase the numbers of pathogenic organisms
   - b. Release ammonia and phosphorus
   - c. Convert BOD into biomass
   - d. Decrease overall particle size

6. Starch can be broken down by bacteria into simple sugars. The starch could be described as
   - a. Hydrolytic
   - b. Biodegradable
   - c. Inert
   - d. Inorganic

7. Which of the following substances can be measured with the COD test, but not the BOD test?
   - a. Biodegradable organics
   - b. Ammonia
   - c. Non-biodegradable organics
   - d. Sulfates

8. Substances that are both soluble and non-biodegradable will be
   - a. Incorporated to biomass
   - b. Adsorbed onto sludge particles
   - c. Outgas as carbon dioxide
   - d. Discharged into the final effluent

9. A sand particle that was not dense enough to be captured by the primary clarifier passes into the biological secondary treatment process. Where is the sand particle most likely to end up once treatment is complete?
   - a. Biomass
   - b. Sludge
   - c. Effluent
   - d. Atmosphere

10. The difference between the soluble COD in the WRRF influent and effluent is
    a. Readily biodegradable COD
    b. Ammonia
    c. Nitrogenous oxygen demand
    d. Isobutyeric acid

Microbiology

1. The cell membrane serves as a selective barrier that only allows some substances to pass through.
    ☐ True
    ☐ False

2. Enzymes are catalysts.
    ☐ True
    ☐ False

3. Protozoans are predators that can consume large quantities of free-swimming bacteria.
    ☐ True
    ☐ False

4. Metazoa are multicellular animals and are typically smaller than both protozoans and bacteria.
    ☐ True
    ☐ False

5. A bacterium can move _____ across its cell membrane without expending energy.
    a. Soluble CBOD
    b. Water
    c. Particulate COD
    d. Phosphate ions

6. Extracellular polymeric substances:
    a. Help the cell adhere to surfaces and other bacteria
    b. Form a protective coating around the cell
    c. Consist of sugars chemically joined together
    d. All of the above.

7. When a bacterium reproduces, how many daughter cells are produced from each parent cell?
    a. 1
    b. 2
    c. 3
    d. 4

8. A single bacterium enters the treatment process and has access to unlimited amounts of food, energy, oxygen, and the other substances it needs to grow and reproduce. If the bacterium can reproduce every 20 minutes, how many bacteria will there be after 1 hour?
    a. 2
    b. 4
    c. 8
    d. 16

9. During the lag phase of bacterial growth:
    a. No growth is observed
    b. Growth rate rapidly increases
    c. Growth rate and death rate are about equal
    d. Death rate exceeds growth rate

Bacteria in Secondary Treatment Processes

1. Sugar is combined with oxygen to produce carbon dioxide and water. In this reaction, the sugar is being oxidized.
   - ☐ True
   - ☐ False

2. Autotrophic bacteria consume CBOD and produce new bacteria, carbon dioxide, and water.
   - ☐ True
   - ☐ False

3. Autotrophic bacteria grow faster than heterotrophic bacteria because their fuel source is energy rich.
   - ☐ True
   - ☐ False

4. The phosphate accumulating organisms must be cycled between anaerobic and anoxic or aerobic conditions to perform luxury uptake of phosphorus.
   - ☐ True
   - ☐ False

5. Heterotrophic bacteria obtain both their energy and their carbon from
   a. Carbonate and bicarbonate ion
   b. DO
   c. Biodegradable organic material
   d. Sulfate

6. Algae are classified as
   a. Autotrophs
   b. Phototrophs
   c. Heterotrophs
   d. Taxitrophs

7. Match the environmental condition to its description.
   1. Anaerobic   a. Dissolved oxygen present
   2. Anoxic   b. Nitrate present, no DO present
   3. Aerobic   c. Neither oxygen nor nitrate present

8. When heterotrophic bacteria consume CBOD in the presence of oxygen, the byproducts will be
   a. Nitrate, nitrite, and water
   b. Nitrate and carbon dioxide
   c. Carbon dioxide and water
   d. Carbon dioxide and nitrogen gas

9. Obligate aerobes require _____ for respiration.
   a. Nitrate
   b. Oxygen
   c. Ammonia
   d. CBOD

10. A heterotrophic bacterium is placed under anoxic conditions. This bacterium will
   a. Convert CBOD to CO_2 and H_2O using oxygen
   b. Convert NH_3^- to NO_2^- and NO_3 using oxygen
   c. Convert CBOD to N_2, CO_2, and H_2O using nitrate
   d. Convert NH_3 to N_2, CO_2, and H_2O using nitrate

11. The process of consuming CBOD with nitrate is called
   a. Denitrification
   b. Nitrification
   c. Fermentation
   d. Hydrolysis

12. **The NOB**
    a. Consume only soluble CBOD
    b. Require an inorganic carbon source
    c. Prefer VFAs
    d. Convert ammonia to nitrite

13. **Phosphate accumulating bacteria obtain the energy they need to pick up VFAs under anaerobic conditions from**
    a. VFAs
    b. Adenosine triphosphate
    c. Polyphosphate
    d. Poly-β-hydroxybutyrate

14. **Sulfate-reducing bacteria produce this toxic gas under anaerobic conditions.**
    a. Carbon dioxide
    b. Nitrogen gas
    c. Hydrogen sulfide
    d. Nitrous oxide

15. **Production of sulfuric acid in collection systems and WRRFs is a concern because**
    a. It is toxic to humans at low concentrations in air.
    b. It corrodes concrete and steel.
    c. It interferes with metabolism of CBOD at low concentrations.
    d. It consumes DO.

Microbial Growth Rates

1. Every type of bacteria has a maximum growth rate that can be achieved when resources are present in excess and environmental conditions are favorable.
   ☐ True
   ☐ False

2. When resources are limited, the growth rate of the bacteria will be proportional to the concentration of the most limited substrate.
   ☐ True
   ☐ False

3. The saturation coefficient for a particular heterotrophic bacteria is 20 mg/L of BOD. The growth rate will increase rapidly at BOD concentrations above 20 mg/L.
   ☐ True
   ☐ False

4. There can be only one limiting substrate at any time.
   ☐ True
   ☐ False

5. The amount of new microorganisms produced each day is dependent on the amount of food entering the treatment process.
   ☐ True
   ☐ False

6. Assuming that a bacterium has all the nutrients and environmental conditions it needs to grow and reproduce, when will growth rates be fastest?
   a. When the water temperature reaches 10 °C
   b. After all of the BOD has been consumed
   c. When the water temperature is 22 °C
   d. Before ammonia concentrations reach 15 mg/L

7. This term is used to describe how fast or slow a chemical or biological reaction takes place.
    a. Monod
    b. Kinetics
    c. Saturation coefficient
    d. Growth

8. A common limiting factor for bacterial growth in secondary treatment processes is
    a. Excess BOD
    b. pH near neutral
    c. Abundant nutrients
    d. Accumulation of toxic byproducts

9. This equation predicts the growth rate of bacteria based on the availability of resources.
    a. Yield
    b. Monod
    c. Decay curve
    d. μ_{max} parametric

10. The influent BOD to a WRRF suddenly decreases by 30%. The number of new bacteria produced in the secondary treatment process
    a. Is dependent on the influent flowrate
    b. Will increase according to the BOD-to-ammonia ratio
    c. Is unaffected by the change
    d. Will decrease to match the influent BOD supply

11. The half-saturation coefficent of BOD for a particular bacteria is 15 mg/L and the maximum growth rate is 12 g VSS/g VSS·d. If the concentration of BOD in the process is 15 mg/L, what is the growth rate of the bacteria?
    a. $6\,d^{-1}$
    b. $12\,d^{-1}$
    c. $15\,d^{-1}$
    d. $24\,d^{-1}$

CHAPTER 6
Wastewater Treatment Ponds

Purpose and Function

1. All wastewater pond systems have a headworks to remove grit and rags before the wastewater enters the pond.
   - ☐ True
   - ☐ False

2. Membrane covers on anaerobic ponds capture methane gas for reuse.
   - ☐ True
   - ☐ False

3. Wastewater pond systems can reliably remove nutrients from their effluent.
   - ☐ True
   - ☐ False

4. This type of pond treatment system is best suited to high-strength industrial waste.
   a. Aerated
   b. Aerobic
   c. Anaerobic
   d. Facultative

5. The most commonly used type of wastewater pond system is
   a. Aerated
   b. Aerobic
   c. Anaerobic
   d. Facultative

6. This type of wastewater pond is shallow enough for sunlight to penetrate the entire pond.
   a. Aerobic
   b. Anoxic
   c. Facultative
   d. Anaerobic

7. How often does an evaporation pond typically discharge effluent to a receiving stream, river, or lake?
   a. Continuously
   b. Weekly
   c. Semiannually
   d. Never

Theory of Operation—Facultative Ponds

1. Most of the dissolved oxygen in unaerated aerobic and facultative ponds comes from
   a. Surface aerators
   b. Wind
   c. Photosynthesis
   d. Respiration

2. These organisms are primarily responsible for BOD removal in wastewater ponds:
   a. Algae
   b. Bacteria
   c. Protozoans
   d. Ducks

3. The first pond in a wastewater pond system is similar to what unit process in a mechanical treatment facility?
   a. Headworks
   b. Primary clarification
   c. Trickling filter
   d. Disinfection

4. The minimum number of ponds required by most regulatory agencies to prevent short-circuiting is
   a. One
   b. Two
   c. Three
   d. Four

5. The main gas produced in any type of wastewater pond system by bacteria is
   a. Carbon dioxide
   b. Methane
   c. Nitrogen
   d. Oxygen

6. A wastewater pond system consists of five ponds operated in parallel. Raw influent wastewater is fed to the first two ponds. How many primary ponds are there?
   a. One
   b. Two
   c. Three
   d. Four

7. Match the organism to its function.

   1. Algae   a. Convert ammonia to nitrite and nitrate
   2. Bacteria   b. Consume BOD and release hydrogen sulfide
   3. AOB and NOB   c. Produce oxygen
   4. Acid formers   d. Convert VFAs into methane
   5. Sulfate-reducing bacteria   e. Convert settleable BOD and biomass into VFAs
   6. Methanogens   f. Consume BOD and release nutrients

8. About half of the organic material that settles to the bottom of the pond will ultimately be converted into
   a. Algae and bacterial biomass
   b. Methane and carbon dioxide
   c. Carbon dioxide and water
   d. Volatile fatty acids and oxygen

Pond Biology

1. The pH in an unaerated, facultative pond will be lowest
   a. Early morning
   b. Before lunch
   c. Late afternoon
   d. At night

2. Algae obtain carbon for growth and reproduction from
   a. BOD
   b. Volatile fatty acids
   c. Dissolved carbon dioxide
   d. Sugar

3. During photosynthesis, algae produce
   a. Nitrogen gas
   b. Oxygen
   c. Carbon dioxide
   d. Methane

4. Dissolved oxygen concentrations will be highest in an unaerated, facultative pond
   a. On a bright, sunny day
   b. After a daphnia bloom
   c. After 3 days of rain
   d. In the middle of the night

5. This type of algae is rarely seen in wastewater treatment ponds:
   a. Brown
   b. Green
   c. Blue-green
   d. Red

6. A healthy pond with high dissolved oxygen should appear
   a. Brownish
   b. Pink to dark purple
   c. Grey to light green
   d. Green and sparkling

7. Ponds tend to turn from green to brown as the days get shorter and less sunlight is available because
   a. Lack of dissolved oxygen prevents algae growth
   b. Diatoms can capture light over more of the light spectrum
   c. Presence of large numbers of daphnia and rotifers
   d. Colder water temperatures

8. A sudden increase in the amount of algae in a pond is called a
   a. Burst
   b. Flowering
   c. Bloom
   d. Sparkle

9. This unique ability of cyanobacteria, also called the blue-green algae, allows them to prosper when green algae cannot:
   a. Use nitrogen gas from the atmosphere
   b. Recycle phosphorus internally
   c. Reproduce faster under ideal conditions
   d. Obtain carbon from BOD

10. Cyanobacteria blooms are most likely to occur in the
    a. Spring
    b. Summer
    c. Autumn
    d. Winter

11. Cyanobacteria are problematic in natural and wastewater pond systems because
    a. They increase phosphorus concentrations
    b. Some species produce toxins
    c. They support daphnia blooms
    d. Ponds may become overoxygenated

12. A small community hosts an annual festival that doubles or triples the town's population. During the festival, the surface of the wastewater pond turns from sparkling green to dark pink. What is the most likely cause?
    a. Increased BOD loading
    b. Growth of cyanobacteria
    c. Flowering of green algae
    d. Portable toilet chemicals

13. Red clouds of tiny, darting organisms are observed at the edge of a pond in the early morning. Which of the following must be true?
    a. This pond recently experienced a bloom of blue-green algae.
    b. Dissolved oxygen concentrations are low.
    c. Shoreline vegetation should be trimmed to remove breeding habitat.
    d. Hydrogen sulfide generation has increased in the sludge blanket.

14. A bloom of daphnia or rotifers is most likely to occur
    a. When dissolved oxygen concentrations are lowest
    b. After a storm event or several overcast days
    c. Immediately after a bloom of green algae
    d. When toxic compounds are present in the influent

15. A facultative pond turns dark purple. Which of the following statements must be true?
    a. Daphnia populations have bloomed in response to a bloom of green algae.
    b. Sludge blanket nitrite and nitrate concentrations have dropped below a critical level.
    c. Anaerobic conditions have caused a bloom of purple sulfur bacteria.
    d. Pond pH and DO are both increasing.

Design Parameters

1. BOD_5 loading to facultative ponds is limited by
    a. Surface area
    b. Available oxygen
    c. Sludge storage volume
    d. Hydraulic detention time

2. A mechanically aerated facultative pond system has three cells operated in series. Dissolved oxygen concentrations in the primary pond have been dropping steadily and are now below 1 mg/L even in the late afternoon. The operator should
    a. Bypass flow to the polishing pond
    b. Decrease aerator run time
    c. Consider switching to parallel operation
    d. Add phosphate to grow green algae

3. Facultative ponds are capable of processing more BOD load when
    a. Wastewater is colder
    b. Less oxygen is provided
    c. Wastewater is warmer
    d. Cyanobacteria are present

Expected Performance

1. Given the following information, what is the most likely cause of high TSS in the final effluent? Pond is located in Minnesota. It is early spring. Air temperatures have increased from 5 °C (41 °F) to 15 °C (59 °F) over the last 5 weeks. The pond is free of ice, but has not yet achieved its summer green sparkle. Effluent BOD_5 is less than 10 mg/L. Effluent TSS is 120 mg/L.
   a. Increased algae production
   b. Short-circuiting
   c. Incomplete treatment
   d. Pond turnover

2. When is green algae growth most likely to contribute to high effluent TSS?
   a. Early spring
   b. Summer
   c. Fall
   d. Winter

3. A facultative pond has a pH of 8.5 and an abundance of green algae. The hydraulic detention time is less than 10 days. The influent ammonia concentration is 25 mg/L as N and the final effluent concentration is 8 mg/L as N. What is the primary method of ammonia removal under these operating conditions?
   a. Assimilative uptake
   b. Conversion to nitrite and nitrate
   c. Volatilization to atmosphere
   d. Particles settling into sludge

4. For biological nitrification to take place in a wastewater pond, which of the following conditions must be met?
   a. Long HRT and warm temperatures
   b. Short HRT and warm temperatures
   c. Short HRT and cold temperatures
   d. Long HRT and cold temperatures

Equipment

1. Calculated detention times and actual detention times are nearly identical for pond systems.
   ☐ True
   ☐ False

2. Dike slopes are steeper on the outside of the pond than on the inside of the pond.
   ☐ True
   ☐ False

3. In general, larger ponds require larger rocks for shore protection.
   ☐ True
   ☐ False

4. An operator notices several large bald patches on the top of a dike where the grass no longer grows. It is almost winter and heavy snows are likely. The operator should
   a. Mow the surrounding grass and reseed.
   b. Place a layer of gravel over the bald areas.
   c. Wait until spring to reseed the grass.
   d. Postpone further dike inspections until spring.

5. Vegetation should be removed from the shoreline and grass kept neatly trimmed to
   a. Maintain the appearance of the facility.
   b. Protect the pond liner and prevent leaks.
   c. Make the pond inhospitable to burrowing animals.
   d. Increase mosquito and midge larvae to feed fish.

6. **Match the term to its definition.**

   1. Freeboard
   2. Riprap
   3. Fetch
   4. Emergent plant
   5. Bentonite
   6. PRV
   7. Impermeable

   a. Plant that grows in shallow water
   b. Clay mineral used for lining ponds
   c. Distance from top of water to top of dike
   d. Water cannot pass through
   e. Distance wind travels over the water
   f. Loose stone placed to prevent erosion
   g. Prevents ballooning of pond liner

7. **Riprap should extend at least _____ m (ft) above and below the minimum and maximum water levels.**
   a. 0.3 m (1 ft)
   b. 0.7 m (2 ft)
   c. 1.0 m (3.3 ft)
   d. 1.3 m (4.3 ft)

8. **When driving vehicles on the top of pond dikes, operators should _____ to minimize erosion.**
   a. Keep tires within established ruts
   b. Drive on wet dikes to recompact soil
   c. Minimize use of gravel on dike sides
   d. Vary driving patterns

9. **Dikes should be inspected at least _____ for evidence of burrowing animals.**
   a. Daily
   b. Weekly
   c. Monthly
   d. Quarterly

10. **The primary reason for lining a pond is to**
    a. Discourage burrowing
    b. Reduce weed growth
    c. Prevent seepage
    d. Eliminate mosquitos

11. **An operator needs to empty a pond for repairs. They calculate the volume of the pond and determine that their dewatering pump should be able to completely empty the pond in 9 days. The water level dropped rapidly in the beginning, but after two weeks, it seems the pump isn't moving any water at all. The water level just won't go below 0.7 m (2 ft). What is the most likely cause?**
    a. Water is passing through the dike from the adjacent pond.
    b. Groundwater is entering the pond through an open PRV.
    c. Area of pond increases with depth, so it appears the pumping rate slowed.
    d. Inlet valve is in the full open position.

12. **A facultative pond has cattails growing up in the middle of the pond. What must be true?**
    a. Mosquito breeding ground has been eliminated.
    b. There are tears in the plastic liner.
    c. The bentonite clay liner has been compromised.
    d. Water depth at that location is less than 1.5 m (5 ft).

13. **This type of liner provides the highest level of protection for groundwater.**
    a. Bentonite
    b. Natural soil
    c. Hot mix asphalt
    d. Synthetic, plastic

14. An inlet diffuser on a pond
    a. Is typically only perforated at one end
    b. Distributes the influent flow across the pond
    c. Requires a riser pipe or T-structure
    d. Would be located at the center of a newly constructed pond

15. A facultative pond has several inlet pipes at different locations and depths. It is winter and the water temperature within the pond is 5 °C (41 °F) and the influent wastewater temperature is 15 °C (59 °F). To minimize the possibility of short-circuiting, which inlet pipe should be used?
    a. Inlet near the center of the pond across from the outlet
    b. Inlet nearest the top of the water surface and furthest from the outlet
    c. Inlet closest to the pond bottom and furthest from the outlet
    d. Inlet at the bottom of the pond and near the center

16. Pond outlet structures often have two pipes put together to form a T-shape to
    a. Prevent settled sludge from being drawn into the pipe
    b. Eliminate thermal stratification of the pond
    c. Reduce short-circuiting by decreasing mixing
    d. Provide operational flexibility to the operator

17. This device is placed around a pipe penetrating a dike to prevent water from traveling from one side of the dike to the other.
    a. Bollard
    b. Bedding wall
    c. PRV
    d. Seepage collar

Process Variables

1. In general, sludge accumulates faster in ponds located in Michigan than in ponds located in Alabama.
    ☐ True
    ☐ False

2. Algae growth causes pond pH to increase during the day.
    ☐ True
    ☐ False

3. The oxygen saturation concentration at 20 °C (68 °F) is lower at sea level than it is in the mountains.
    ☐ True
    ☐ False

4. A pond system with three cells is currently operated in series with one primary pond. Over the last few weeks, temperatures have decreased and the pond has turned from green to brown. The operator notices red streaks at the surface of the primary pond that persist through late afternoon. What corrective action should the operator take to eliminate the red streaks?
    a. Add diatoms to consume daphnia.
    b. Lower the pond level in the primary pond.
    c. Switch from series to parallel operation.
    d. Increase loading to the primary pond.

5. The influent flow to a primary pond contains 250 mg/L of BOD_5. The influent flow rate is 5000 m³/d (1.32 mgd). How many kilograms (pounds) per day of BOD_5 are entering the primary pond?
    a. 149 kg/d (330 lb/d)
    b. 1120 kg/d (2468 lb/d)
    c. 1250 kg/d (2752 lb/d)
    d. 1650 kg/d (3637 lb/d)

6. A wastewater pond system has four ponds that are all the same size. The design criteria for the facility state that the maximum organic loading rate to the primary pond should be kept below 40 kg/ha·d (36 lb/ac/d). If all of the influent BOD$_5$ load is sent to one pond, the loading rate will be 65 kg/ha·d (58.5 lb/ac/d). How many ponds should receive raw influent wastewater to keep the system within its design parameters?

    a. 1
    b. 2
    c. 3
    d. 4

7. The dissolved oxygen concentration will be the highest in a facultative pond when

    a. The water is cold and sunlight is plentiful
    b. The water is warm and all mechanical aerators are operating
    c. The water is cold and all mechanical aerators are operating
    d. The water is warm and sunlight is plentiful

8. A primary pond has a surface area of 18.75 ha (46.3 ac). Find the loading rate when the influent flow is 6000 m³/d (1.6 mgd) and the BOD concentration is 180 mg/L.

    a. 28.8 kg/ha·d (26.0 lb/ac/d)
    b. 57.6 kg/ha·d (51.9 lb/ac/d)
    c. 115.2 kg/ha·d (103.8 lb/ac/d)
    d. 172.8 kg/ha·d (155.7 lb/ac/d)

9. This chemical may be used to decrease effluent pH prior to discharge.

    a. Sodium hydroxide (NaOH)
    b. Calcium carbonate (CaCO$_3$)
    c. Sodium bicarbonate (NaHCO$_3$)
    d. Sulfuric acid (H$_2$SO$_4$)

10. Sludge accumulation in a wastewater pond system will be greatest at the

    a. Inlet to primary pond
    b. Outlet of primary pond
    c. Inlet to settling pond
    d. Outlet of settling pond

11. Sludge blanket depth should be checked

    a. Monthly
    b. Quarterly
    c. Annually
    d. Every 5 years

Operation

1. Dyes that block sunlight are approved by U.S. EPA for use in wastewater ponds to control algae.
    ☐ True
    ☐ False

2. Sludge blanket depth should be monitored at multiple locations throughout all ponds.
    ☐ True
    ☐ False

3. Algae provide oxygen to aerobic and facultative ponds and

    a. Consume organic matter
    b. Contribute to effluent TSS
    c. Can be killed with barley straw
    d. Grow symbiotically with duckweed

4. A three-pond system is in parallel operation, but dissolved oxygen concentrations in the primary cells are still too low. If equipment is available, the oxygen concentration could be increased by
    a. Reducing surface aerator speed
    b. Increasing recycle flow from the settling pond
    c. Adding barley straw booms
    d. Decreasing surface aerator run time

5. Duckweed can outcompete algae for
    a. Nitrogen
    b. Phosphorus
    c. Oxygen
    d. Sunlight

6. Which of the following algae control methods is an algistat?
    a. Barley straw
    b. Ultrasound
    c. Daphnia
    d. Rotifers

Maintenance

1. An example of preventative maintenance is
    a. Replacing a broken drive belt
    b. Repairing pond dikes
    c. Routine lubrication of motors
    d. Replacing a seized motor

2. Valves and gates should be exercised
    a. By completely opening and closing them several times
    b. During quarterly inspections for surface rust
    c. Any time leakage is observed around the stem
    d. Immediately before placing a pond into service

3. Burrowing animals must be removed
    a. As soon as they are noticed
    b. Prior to breeding season
    c. Only if dikes and liners are impacted
    d. In early spring before the pond thaws

4. Before using any herbicide to control vegetation at a WRRF
    a. Plants should be cut as short as possible to increase effectiveness.
    b. Burrowing animals should be relocated to other areas.
    c. A test application should be tried on one-quarter of the pond.
    d. Operators should contact their regulatory agency for guidance.

5. Once a regulatory agency has approved an herbicide for use, it may be used again in the future.
    ☐ True
    ☐ False

6. An effective method for resettling floating sludge is to spray it with a high pressure hose.
    ☐ True
    ☐ False

CHAPTER 7
Fixed-Film Treatment

Purpose and Function of Fixed-Film Processes

1. In fixed-film processes, the microorganisms grow attached to some type of support media.
   - ☐ True
   - ☐ False

2. Some fixed-film processes convert ammonia to nitrite and nitrate.
   - ☐ True
   - ☐ False

3. For a trickling filter to function properly, all of the media must be submerged.
   - ☐ True
   - ☐ False

4. Fixed-film processes don't typically remove _____ beyond assimilative uptake.
   a. BOD
   b. TSS
   c. Nitrogen
   d. Phosphorus

5. Excess biofilm leaves a fixed-film process by
   a. Mechanical wasting
   b. Scraping
   c. Sloughing
   d. Power washing

6. The primary difference between trickling filters and RBCs is
   a. RBCs rotate through the wastewater.
   b. Only trickling filters can remove ammonia.
   c. Sloughing rates are lower in RBCs.
   d. Trickling filters do not have influent pumps.

Theory of Fixed-Film Operation

1. Only soluble wastewater components can diffuse into the biofilm and be utilized by the microorganisms.
   - ☐ True
   - ☐ False

2. In natural systems, substances move from areas of low concentration to areas of high concentration.
   - ☐ True
   - ☐ False

3. The amount of EPS present in biofilms decreases with biofilm age.
   - ☐ True
   - ☐ False

4. The first step in biofilm formation is
    a. Attachment
    b. Initial Growth
    c. Colony Formation
    d. Sloughing

5. Microorganisms in the biofilm excrete _____ to help convert particulate BOD into soluble BOD.
    a. Exopolymeric substances
    b. Enzymes
    c. Poisons
    d. Digestive acids

6. Which of the following substances can diffuse through a biofilm?
    a. Total suspended solids
    b. Colloidal BOD
    c. Dissolved oxygen
    d. Food particles

7. Microorganisms in this portion of the biofilm have the greatest access to resources and will grow the fastest.
    a. Outermost layer
    b. Anoxic layer
    c. Anaerobic layer
    d. Media surface

8. This substance, secreted by bacteria, helps the biofilm attach to the supporting media.
    a. Enzymes
    b. Exopolymer
    c. Carbon dioxide
    d. Lipocolonic glucosamine

9. A fixed-film process receives a heavy organic load. Which of the following statements is true?
    a. Oxygen will easily penetrate the full depth of the biofilm.
    b. The biofilm will contain high numbers of nitrifying bacteria.
    c. Growth will be limited due a lack of soluble BOD.
    d. Heterotrophic bacteria will overgrow the nitrifying bacteria.

10. Sloughed biofilm from a trickling filter or RBC
    a. Is recycled to the front of the treatment process
    b. Is removed by gravity in a secondary clarifier or hopper
    c. Is typically less dense than the surrounding wastewater
    d. Consists primarily of live microorganisms

11. Filter flies require this condition for egg laying and reproduction.
    a. Low organic loading rates
    b. High flushing intensities
    c. DO greater than 2 mg/L
    d. Dry surfaces

12. Which of the following statements about snail infestations is NOT true?
    a. Consume large quantities of biofilm
    b. May contribute to effluent BOD
    c. Increase TSS removal
    d. Shells can damage equipment

13. These sessile organisms can be found in the biofilms of lightly loaded trickling filters.
    a. Filter flies
    b. Snails
    c. Worms
    d. Rotifers

Trickling Filters

1. Trickling filters are classified according to their organic loading rate.
    ☐ True
    ☐ False

2. An intermediate rate trickling filter may remove both BOD and ammonia when lightly loaded.
    ☐ True
    ☐ False

3. As loading rates to trickling filters increase, the percent removal also increases.
    ☐ True
    ☐ False

4. The organic loading rate to a trickling filter is currently near 1.5 kg/m³·d (100 lb BOD_5/d/1000 cu ft). What is the most likely effect of reducing the loading rate to 0.07 kg/m³·d (5 lb BOD_5/d/1000 cu ft)?
    a. Effluent TSS will increase
    b. Soluble BOD will pass through
    c. Effluent ammonia will decrease
    d. Biofilm growth will increase

5. Trickling filters are capable of removing all of the following wastewater components EXCEPT:
    a. CBOD
    b. TSS
    c. Fecal coliforms
    d. Ammonia

6. Treated wastewater is often recycled back to the front of a trickling filter process to
    a. Increase the organic loading rate
    b. Control biofilm thickness
    c. Decrease wetting rates
    d. Capture sloughed biofilm

7. One advantage of plastic media over rock media is
    a. Allows higher loading rates
    b. Locally available building material
    c. Shallower media beds
    d. Improved TSS removal efficiency

8. Two-stage, alternating double filtration trickling filters processes regularly switch which filter receives primary effluent for this reason.
    a. Ensure adequate removal of phosphorus
    b. Promote uniform biofilm growth in both filters
    c. Improve hydraulic wetting rates
    d. Decrease biofilm sloughing events

9. A tertiary trickling filter, placed after the secondary clarifier, is intended to remove
    a. BOD
    b. TSS
    c. Ammonia
    d. Nitrate

Trickling Filter Equipment

1. All trickling filters are equipped with motors, located at the top of the inlet column, to rotate the distributor arm.
    - ☐ True
    - ☐ False

2. In a naturally ventilated trickling filter, the direction of airflow may change during the day.
    - ☐ True
    - ☐ False

3. Orifices on the leading edge of a trickling filter arm
    a. May be used to reduce arm speed
    b. Help push the distributor arm around
    c. Should all be in the full open position
    d. Rinse the outer wall of the trickling filter

4. The underdrain system of a trickling filter
    a. Helps distribute wastewater evenly over the media
    b. Must be cleaned daily to prevent plugging
    c. Collects treated wastewater and aids in ventilation
    d. Slopes away from the underdrain channel

5. The open space below the underdrain system in a plastic media trickling filter is called the
    a. Channel
    b. Ventilator
    c. Stanchion
    d. Plenum

6. This device is used to level the distributor arms of a trickling filter.
    a. Stanchion
    b. Turnbuckle
    c. Shims
    d. Dump gate

Biofilm Support Media

1. Match the media type to its defining characteristic.
    a. Rock   1. Also called dump media
    b. Redwood   2. Used in roughing filters
    c. Random   3. Low specific area to volume ratio
    d. Vertical-Flow   4. Rotated sheets mix and aerate
    e. Cross-Flow   5. Narrow slats like pallets

2. Which type of media has the highest SSA?
    a. River Rock
    b. Redwood
    c. Cross-Flow
    d. Random Pack

3. An advantage of rock media is
    a. Larger void spaces than random media
    b. Light weight facilitates taller towers
    c. Potential for blinding
    d. Inexpensive and locally available

4. Voids in trickling filter media
    a. Should be avoided in most circumstances
    b. Allow the flow of wastewater and air
    c. Are responsible for short-circuiting
    d. Consume less than 25% of the total volume

5. As the SSA of trickling filter media increases
    a. More biofilm can be grown in a smaller volume.
    b. The likelihood of blockages decreases.
    c. BOD and ammonia loading rates must decrease.
    d. TSS removal efficiency increases.

Containment Tanks and Structures

1. Ventilation ports are typically located around the perimeter of the trickling filter and below the underdrain system.
    ☐ True
    ☐ False

2. Bearings located at the top of the mast are more likely to corrode than bearings located in the base of the center column.
    ☐ True
    ☐ False

3. The outer walls of a trickling filter often extend 1.2 to 1.5 m (4 to 5 ft) above the top of the media bed for this reason:
    a. Prevent maintenance workers from falling
    b. Protects the distributor from interference by wind
    c. Provides a staging area for new biofilm growth
    d. Prevents wind from blowing media out of structure

4. Plastic media trickling filters are often covered with domes to
    a. Maintain humidity in the plenum
    b. Encourage the growth of algae
    c. Reduce oxygen transfer
    d. Protect media from UV light degradation.

5. The louvers on trickling filter ports
    a. May be partially closed during cold weather to decrease airflow
    b. Require weekly exercising to prevent them from rusting shut
    c. Are only used with mechanically ventilated trickling filters
    d. Must be closed prior to flooding the trickling filter

6. Most hydraulically driven rotary distributors on trickling filters rotate at this rate:
    a. 1 rpm
    b. 3 rpm
    c. 5 rpm
    d. 10 rpm

7. Splash plates on a rotary distributor
    a. Must be replaced quarterly
    b. Operate in the fully open or fully closed position
    c. Prevent trash and debris from clogging the filter
    d. Distribute wastewater over the media

8. One disadvantage of hydraulically driven rotary distributors on trickling filters is
    a. Low energy usage and cost
    b. Orifices are more prone to blockage
    c. High winds may slow or stop the rotation
    d. Slower rotation speeds impede biofilm growth

9. Instability and oscillations of a rotary distributor may occur if
    a. Arms are not kept level and balanced.
    b. Biofilm distribution on media is uneven.
    c. Wastewater flows are split evenly between arms.
    d. Roller bearing plate becomes worn.

Process Variables for Trickling Filters

1. The organic loading rate to a trickling filter is the mass of BOD_5 per square meter (square foot) of surface area.
    ☐ True
    ☐ False

2. Trickling filters may be operated in parallel or in series.
    ☐ True
    ☐ False

3. When the diameter of a pipeline is doubled, the cross-sectional area
    a. Increases by 50%
    b. Increases by 100%
    c. Increases by 200%
    d. Increases by 400%

4. One benefit of increasing recycle flows during times of low primary effluent flow followed by decreasing recirculation flows when primary effluent flows increase is
    a. Maintains constant HLR
    b. Increases biofilm thickness
    c. Maintains constant organic loading rate
    d. Reduced pumping costs

5. When the rotational speed of the distributor is increased
    a. Biofilm thickness increases.
    b. The SK value increases.
    c. The HLR will match the OLR.
    d. Recycle flows must decrease.

6. All of the following methods may be used to increase the flushing intensity for a trickling filter EXCEPT:
    a. Close all of the orifices on two arms of a four-arm rotary distributor.
    b. Decrease the rotational speed by opening orifices on the leading edge of the arms.
    c. Increase the recirculation ratio to increase the total HLR.
    d. Place an additional trickling filter into service to increase surface area.

7. Find the surface area of a trickling filter that is 24.4 m (80 ft) in diameter and 1.8-m (6-ft) tall.
    a. 116.8 m² (1256 cu ft)
    b. 210.2 m² (7536 cu ft)
    c. 467.4 m² (5024 cu ft)
    d. 841.3 m² (30 144 cu ft)

8. The primary effluent flow is 5.7 ML/d (1.5 mgd) and the BOD concentration is 120 mg/L. Find the total mass of BOD in kilograms (pounds) going to the trickling filter when a recycle ratio of 1.5 is used.
    a. 684 kg (1501.2 lb)
    b. 1020 kg (2251.8 lb)
    c. 1368 kg (3002.4 lb)
    d. 1704 kg (3753 lb)

9. The primary effluent flowrate is 30.3 ML/d (8 mgd). The operator would like to increase the total flow to the trickling filter to 60.6 ML/d (16 mgd). What recycle ratio is needed to reach the desired total flowrate?

    a. 0.5

    b. 1

    c. 1.5

    d. 2

10. A trickling filter that is 41.1 m (135 ft) in diameter and 7.3-m (24-ft) deep receives 22.7 ML/d (6 mgd) of primary effluent. The influent BOD concentration is 180 mg/L and the recycle ratio is 2. Find the organic loading rate to the trickling filter in kg/m³·d (lb BOD/d/1000 cu ft).

    a. 0.11 kg/ m³·d (6.6 lb BOD_5/d/1000 cu ft)

    b. 0.42 kg/ m³·d (26.2 lb BOD_5/d/1000 cu ft)

    c. 0.84 kg/ m³·d (52.5 lb BOD_5/d/1000 cu ft)

    d. 1.68 kg/ m³·d (104.9 lb BOD_5/d/1000 cu ft)

11. Calculate the HLR to a trickling filter given the following information. Surface area is 1326 m² (14 306.625 sq ft). Influent flow is 30.3 ML/d (8 mgd) and the recycle ratio is 2.

    a. 22.9 m³/m²·d (559.2 gpd/sq ft)

    b. 34.2 m³/m²·d (838.8 gpd/sq ft)

    c. 68.5 m³/m²·d (1677.5 gpd/sq ft)

    d. 272.6 m³/m²·d (6710.2 gpd/sq ft)

12. Find the SK value for a 24.4-m (80-ft) diameter trickling filter receiving a total flow of 117 ML/d (31 mgd). The filter has a rotary distributor with 4 arms. The distributor completes .3 revolutions per minute.

    a. 36.19 mm/pass (1.4 in./pass)

    b. 144.9 mm/pass (5.7 in./pass)

    c. 289.52 mm/pass (11.4 in./pass)

    d. 579.04 mm/pass (22.8 in./pass)

Ventilation

1. Water entering a trickling filter has a temperature of 20 °C (68 °F). Air will rise through the filter on a hot summer day when the air temperature is 29 °C (85 °F).

    ☐ True

    ☐ False

2. Mechanical ventilation uses blowers to force air through the trickling filter media.

    ☐ True

    ☐ False

3. If the biofilm is allowed to become too thick, it will impede airflow through the filter.

    ☐ True

    ☐ False

4. During extreme cold weather conditions, this should be done to reduce heat loss from a trickling filter.

    a. Increase recycle flows.

    b. Close 25 to 50% of air inlets.

    c. Heat the primary effluent.

    d. Double airflow.

5. If some of the voids in trickling filter media are blocked by debris or excessive biofilm growth, this condition may result in

    a. Dosing

    b. Ponding

    c. Abrasion

    d. Flooding

6. This type of trickling filter media is least susceptible to plugging.
   a. Rock
   b. Redwood
   c. Random
   d. Vertical-flow

7. This may be used to remove excess biofilm from trickling filter media.
   a. Sodium pentothal
   b. Sodium hypochlorite
   c. Sodium chloride
   d. Hydrochloric acid

Operation of Trickling Filters

1. One indication that the recirculation ratio is within the right range is an effluent DO concentration between 3 and 8 mg/L.
   ☐ True
   ☐ False

2. Taking grit basins and primary clarifiers out of service for several weeks will have no effect on trickling filter operation.
   ☐ True
   ☐ False

3. Biofilm generally grows faster at higher temperatures.
   ☐ True
   ☐ False

4. If the results from the pan test differ by more than 5%,
   a. Orifices on the leading edge should be closed.
   b. The stay rods may need to be adjusted.
   c. Recycle ratio should be increased.
   d. Influent water temperature is too high.

5. After making adjustments to the recycle ratio or distributor speed,
   a. Calculate the HLR and SK value.
   b. Verify proper placement of the splash plates.
   c. Measure biofilm thickness with calipers.
   d. Decrease the effluent chlorine concentration.

6. When placing a trickling filter into service, pumps should be started at a lower pump rate because
   a. Media may be crushed by sudden increase in flow.
   b. Debris could be forced into the distributor nozzles.
   c. Avoids emptying the pumping station wet well.
   d. Prevents excessive stress on column bearings.

7. How long does it take for biofilm to fully develop and provide treatment after a trickling filter is placed into service?
   a. 24 hours
   b. 1 to 3 days
   c. Several weeks
   d. Months

8. When removing a trickling filter from service,
   a. Gradually decrease pumping rates before stopping the pumps.
   b. Use your hands to slow and stop the distributor.
   c. Notify your regulatory agency.
   d. Add chlorine to the distributor arms.

Data Collection, Sampling, and Analysis for Trickling Filters

1. How often should the condensate be drained from oil-lubricated bearings in a rotary distributor?
    a. Daily
    b. Weekly
    c. Monthly
    d. Quarterly

2. Last week, the distributor made 1 complete revolution in 55 seconds. This week, the same distributor is making 1 complete revolution every 70 seconds. The HLR has not changed. What should the operator check?
    a. Oil levels
    b. Guy wires
    c. Bearings
    d. Dump gates

3. Distributor orifices should be cleaned weekly by
    a. Removing each splash plate
    b. Spraying with high-pressure water
    c. Soaking in chlorine solution
    d. Flooding the distributor arms

4. Ice buildup on the distributor arms should be removed as quickly as possible to avoid:
    a. Uneven loads and bearing stress
    b. Condensate accumulation in the oil
    c. Water ponding up inside the arms
    d. Snapping the guy wires

5. Debris should be removed from the top of the trickling filter daily to
    a. Keep the operators busy
    b. Decrease solids loading to the filter
    c. Ensure it does not get stuck in nozzles
    d. Prevent blockages and ponding

6. Maintaining a neat and clean landscape by weeding and keeping grass short
    a. Increases the likelihood that windblown debris will enter the filter
    b. Enhances natural draft through the trickling filter
    c. Helps prevent filter flies and other insect infestations
    d. Extends the life of concrete and fiberglass

7. Oil samples should be collected by draining from the bottom of the oil reservoir because
    a. If water is present, it will be at the bottom.
    b. Pulling from the top can entrain air in the sample.
    c. Sample collected from the top may not be well-mixed.
    d. The oil reservoir will be mostly empty during normal operation.

Rotating Biological Contactors

1. What percentage of an RBC is typically submerged during normal operation?
    a. 20%
    b. 40%
    c. 60%
    d. 80%

2. For an RBC system that uses cosettling, the solids that settle in the secondary clarifier are
    a. Sent directly to solids handling
    b. Typically greater than 10% total solids
    c. Sent to the primary clarifier
    d. Applied to the RBC in stage 1

3. An RBC system has two treatment trains with four stages each. Each stage contains one RBC unit. How many units are there altogether?
    a. 2
    b. 4
    c. 8
    d. 16

4. Loading rates are based on the mass of BOD (kilograms or pounds) per
    a. Square meters or square feet of media
    b. Number of RBCs per train
    c. Cubic meters of 1000 cubic feet of media
    d. Population equivalent

5. If the organic loading rate to the first stage of treatment in an RBC is too high,
    a. Treatment efficiencies will increase.
    b. Biofilm development will be inhibited.
    c. Mass sloughing may occur.
    d. Oxygen concentrations will decrease.

6. An operator notices large patches of white biofilm on an RBC in the first stage of a treatment train. This indicates
    a. Snail predation
    b. Lack of oxygen
    c. Organic underloading
    d. Nitrification

7. An RBC that is lightly loaded
    a. May remove both BOD_5 and NH_3-N
    b. Requires a longer HRT than a highly loaded RBC
    c. Produces poor quality effluent high in BOD_5
    d. Performs better at colder water temperatures

Equipment for Rotating Biological Contactors

1. An RBC treatment train may contain different media densities across the train.
    ☐ True
    ☐ False

2. Metals will contract slightly when heated and expand slightly when cooled. This change in length with temperature is why at least one RBC bearing must accommodate temperature changes.
    ☐ True
    ☐ False

3. Either a motor or air may be used to rotate RBCs.
    ☐ True
    ☐ False

4. Baffles between RBC units
    a. Reduce hydraulic detention time
    b. Separate treatment stages
    c. Cannot be moved
    d. Are always made from concrete

5. Removing the baffles between the first and second sequencing batch reactor (SBR) units
   a. Discourages hydraulic short-circuiting
   b. Promotes full nitrification in stage 2
   c. Distributes the influent load
   d. Decreases oxygen availability

6. RBCs are typically covered to
   a. Prevent algae growth
   b. Reduce losses of volatile organics
   c. Keep out vandals
   d. Protect the distributor arm

7. The amount of media surface area per volume is the
   a. Weighted ratio allowance
   b. SSA
   c. Turndown ratio
   d. Media density factor

8. The individual plastic sheets of media in an RBC are corrugated for this reason:
   a. Increases flexibility of modules
   b. Reduces opportunities for biofilm bridging
   c. Increases available surface area
   d. Decreases oxygen transfer

9. Air-driven RBCs
   a. Rotate 4 to 6 times per minute
   b. Capture air bubbles in cups
   c. Rotate faster than motor-driven units
   d. Have thicker biofilm than motor-driven units

10. One operational problem associated with air-driven RBCs is
    a. Harmonics
    b. Shaft stress
    c. Loping
    d. Shedding

11. This device is used to measure the total weight on the shaft.
    a. Load cell
    b. Strain hoist
    c. Load bearing
    d. Tension cable

Process Variables for RBCs

1. The OLR to an RBC is the mass of BOD_5 divided by the volume of media.
   ☐ True
   ☐ False

2. Ammonia removal requires less HDT than BOD_5 removal.
   ☐ True
   ☐ False

3. The OLR that can be treated in an RBC is limited by
   a. Influent flow
   b. Oxygen availability
   c. Rotation speed
   d. *Beggiatoa*

4. Because RBCs don't effectively treat particulates, this parameter is often used to calculate percent removal.
    a. CBOD
    b. BOD
    c. sBOD
    d. COD

5. An RBC process contains two treatment trains with four RBC units each. Each unit contains 13 935 m² (150 000 sq ft) of media surface area. Find the organic loading rate if 680 kg (1501 lb) of soluble BOD_5 is applied to the entire process in a 24-hour period.
    a. 5.2 g sBOD/m²·d (1.0 lb sBOD/d/1000 sq ft)
    b. 6.0 g sBOD/m²·d (1.3 lb sBOD/d/1000 sq ft)
    c. 19.5 g sBOD/m²·d (4.2 lb sBOD/d/1000 sq ft)
    d. 23.6 g sBOD/m²·d (4.8 lb sBOD/d/1000 sq ft)

6. The influent total BOD_5 concentration to an RBC is 80 mg/L. The effluent soluble BOD_5 is 8 mg/L. What is the percent removal?
    a. 10%
    b. 72%
    c. 90%
    d. 95%

7. The rotational speed of an RBC is increased slightly. Which of the following statements is true?
    a. Biofilm thickness will increase
    b. Oxygen concentration will increase
    c. Bearing temperature will decrease
    d. HLR will increase

Process Control for Rotating Biological Contactors

1. The biomass in the first stage of an RBC treatment process is thick and shaggy. This may indicate
    a. High rotation speed
    b. Organic overloading
    c. Insufficient aeration
    d. Septic conditions

2. Healthy RBC biomass in the first stage of a process removing BOD_5 should be
    a. Gray to black
    b. Red to gold
    c. Light brown
    d. White

3. Load cell results are 10% higher than the rated shaft capacity. The operator should
    a. Reduce rotational speed
    b. Manually remove excess biofilm
    c. Increase organic loading
    d. Seed the system with snails

4. All of the following methods may be used to reduce the organic loading rate in the first stage of an RBC process EXCEPT
    a. Remove baffles between stages
    b. Step-feed primary effluent to later stages
    c. Recirculate RBC effluent to stage 1
    d. Add chlorine to the influent channel

5. An RBC treatment train consists of four stages. The DO concentration leaving the last stage is below 1 mg/L. The operator should
    a. Increase the number of RBC stages and trains in service
    b. Replace any missing baffles between stages
    c. Decrease RBC rotational speed
    d. Resuspend settled solids in the RBC tank

6. Low DO conditions may encourage the growth of this nuisance organism.
    a. Snails
    b. Worms
    c. *Beggiatoa*
    d. Rotifers

Operation of Rotating Biological Contactors

1. Sudden, excessive sloughing of biofilm from a trickling filter or RBC may indicate
    a. BOD$_5$ increase
    b. Toxicity
    c. Snail infestation
    d. Excessive aeration

2. Rotating biological contactor rotation should be monitored per quarter turn to determine if _____ is occurring.
    a. Sloughing
    b. Chain slippage
    c. Loping
    d. Vibration

3. If an RBC must be stopped for a period of time while remaining in a full tank,
    a. Manually rotate the unit one-quarter turn every 4 to 6 hours.
    b. Use a power washer to hydraulically scrub all biofilm from the media.
    c. Immediately shut off the blower and diffused air headers to prevent clogging.
    d. Open the cover doors to increase airflow over the exposed media surface.

4. Allowing an RBC to sit idle in a tank full of wastewater
    a. Is acceptable if idle time is less than 8 hours
    b. Controls the growth of *Beggiatoa* and *Thiothrix*
    c. Is part of a feast and famine control strategy
    d. May result in uneven biofilm growth

5. Dissolved oxygen concentrations should be measured at the end of each stage of treatment in an RBC treatment train
    a. Daily
    b. Weekly
    c. Monthly
    d. Quarterly

6. Excessive accumulation of biofilm on an RBC may cause
    a. Media delamination
    b. Increased rotation speed
    c. Shaft deflection
    d. Lower load cell readings

7. Air-drive systems should be operated with higher airflow rates some percentage of the time to
    a. Remove settled solids from diffusers
    b. Scour excess biofilm from the media
    c. Increase media buoyancy
    d. Reduce blower strain from low airflow

CHAPTER 8
Activated Sludge

Purpose and Function

1. The floc particles that make up the MLSS include both live and dead microorganisms.
   - ☐ True
   - ☐ False

2. The goal of activated sludge process control is to produce an MLSS that flocculates, settles, compacts, and meets the requirements of the discharge permit.
   - ☐ True
   - ☐ False

3. Soluble biochemical oxygen demand (BOD) is removed by _____.
   - a. Bioflocculation
   - b. Conversion into MLSS
   - c. Settling in the clarifier
   - d. Offgassing

4. Which of the following components cannot be removed through bioflocculation?
   - a. Non-biodegradable organic solids
   - b. Biochemical oxygen demand
   - c. Toilet paper fibers
   - d. Ammonia-nitrogen

5. Waste activated sludge
   - a. Is the returned settled sludge to the activated sludge basin
   - b. Allows floc particle size control by breaking up larger floc
   - c. Is the excess sludge removed from the activated sludge process
   - d. Converts large particles into food for the bacteria

6. The purpose of the secondary clarifier is to
   - a. Separate the solids from the treated wastewater
   - b. Promote denitrification
   - c. Manipulate the MLSS concentration
   - d. Reduce soluble BOD

Theory of Operation

1. Floc particles may be kept in suspension in the activated sludge process with air or mixers.
   - ☐ True
   - ☐ False

2. As the MLSS settles, larger particles settle faster than smaller particles, reaching the bottom of the settleometer first.
   - ☐ True
   - ☐ False

3. The organic fraction of activated sludge floc is referred to as
   - a. MLSS
   - b. Bioflocculated TSS
   - c. MLVSS
   - d. Bioslurry volatile fraction

4. The fundamental difference between a wastewater treatment pond system and activated sludge is
    a. Activated sludge systems discharge to polishing ponds.
    b. Pond systems contain larger populations of microorganisms.
    c. Only pond systems remove both BOD_5 and TSS.
    d. Activated sludge systems have a RAS line.

5. Sludge age refers to
    a. The length of time the solids remain in the activated sludge process
    b. How many years an activated sludge process has been in operation
    c. The ratio of the return to WAS pumping rates
    d. How often the return activated sludge pumping rate is adjusted each week

6. One effect of returning sludge to the beginning of the activated sludge process is
    a. Increases total time required to complete treatment
    b. Increases concentration of microorganisms in the process
    c. Links hydraulic and SRT to increase efficiency
    d. Controls the system sludge age by removing excess biomass

7. Anaerobic and anoxic zones use _____ to keep the MLSS in suspension.
    a. Fine bubble diffusers
    b. Mixers
    c. Draft tubes
    d. Recycle pumps

8. One reason air is added to some activated sludge basins is to
    a. Prevent freezing
    b. Keep the MLSS in suspension
    c. Circulate waste products
    d. Maintain water temperature

9. The settled solids in the bottom of the clarifier are called the
    a. Blanket
    b. Septic layer
    c. Schmutzdecke
    d. Scum

10. What is the most likely outcome if excess solids are not removed from the activated sludge basin on a regular basis?
    a. Treatment efficiency will increase.
    b. Sludge settleability will improve.
    c. Solids may be lost to the final effluent.
    d. Oxygen use will decrease.

11. The purpose of the RAS line is to
    a. Transfer settled solids to solids handling processes
    b. Return settled solids to the activated sludge basin
    c. Manipulate MLSS concentrations
    d. Control sludge age

Activated Sludge Microbiology

1. Ideally, filament-forming bacteria will outnumber floc-forming bacteria.
    ☐ True
    ☐ False

2. Match the following terms to their definitions.

    a. Heterotroph   1. Utilizes inorganic carbon
    b. Autotroph   2. Uses DO, nitrite, or nitrate
    c. Aerobic   3. Grows in clumps
    d. Facultative   4. Utilizes organic carbon
    e. Anaerobic   5. Grows in long chains
    f. Floc former   6. Cannot use DO, nitrite, or nitrate
    g. Filament   7. Utilizes DO

3. While some filamentous bacteria provide structure and support to floc particles, too many filaments

    a. Will increase floc particle settling velocity
    b. Require excessive quantities of nutrients
    c. Can cause settling problems in the clarifier
    d. Increase floc density and oxygen usage

4. Which of the following filaments indicates a low DO condition?

    a. *M. parvicella*
    b. Nocardioforms
    c. *N. limicola*
    d. *H. hydrossis*

5. Match these commonly occurring filaments to their preferred environmental condition.

    a. Nocardioforms   1. Septicity
    b. *S. natans*   2. High sludge age
    c. *Beggiatoa*   3. FOG
    d. *M. parvicella*   4. Nutrient deficiency
    e. Type 021N   5. Low DO

6. The growth of the filament *M. parvicella* has been linked to influent FOG. Even if FOG is present, this must also be true for *M. parvicella* to accumulate in the process.

    a. Minimum time to reproduce has been met
    b. Dissolved oxygen concentration is low
    c. Hydrogen sulfide is present in excess
    d. pH is less than 6.5

Design Parameters

1. Compared to high-rate activated sludge processes, low-rate processes

    a. Typically have higher organic loading rates
    b. Typically have longer sludge ages
    c. Remove less ammonia than high-rate systems
    d. Are often less stable than conventional systems

2. Activated sludge processes are characterized by

    a. Sludge age, loading rate, and F/M
    b. Presences of flagellates, ciliates, and rotifers
    c. Percentage removal of ammonia, nitrite, and nitrate
    d. Average MLVSS concentration

3. Surface overflow rate in a secondary clarifier is an example of

    a. Downward velocity
    b. Hydraulic capacity
    c. Loading capacity
    d. Sludge settleability

4. If the solids loading capacity of a secondary clarifier is exceeded

   a. The hydraulic loading rate may be increased to compensate.

   b. SOR must be kept below 100 kg/m²·d (20 lb/d/sq ft).

   c. The clarifier may not be able to settle the incoming MLSS.

   d. The effluent suspended solids concentration should decrease.

5. The BOD_5 concentration in the final effluent from an activated sludge process (after clarification) should be

   a. Less than or equal to 30 mg/L

   b. About 50% of the influent BOD_5 concentration

   c. Directly proportional to the ammonia concentration

   d. Independent of operational mode

Equipment

1. Water resource recovery facilities typically have multiple activated sludge basins and clarifiers so they can be taken out of service for cleaning and maintenance.

   ☐ True

   ☐ False

2. Blowers used with activated sludge processes are low-pressure.

   ☐ True

   ☐ False

3. Having air bubbles in water is the same as having DO.

   ☐ True

   ☐ False

4. Water moves through this type of activated sludge basin in the same way that water moves through a pipe.

   a. Complete mix

   b. Oxidation ditch

   c. Sequencing batch reactor

   d. Plug flow

5. One advantage of complete-mix-type activated sludge systems is

   a. More complicated operation

   b. Dilution of shock loads

   c. Prone to filamentous bulking

   d. Lower oxygen demand

6. Selectors are useful for

   a. Diverting flow between basins

   b. Thickening WAS

   c. Controlling the growth of some filaments

   d. Decreasing total treatment time

7. The holes in the bottom of a peripheral feed clarifier channel

   a. Prevent short-circuiting of MLSS to the weirs

   b. Must have their o-rings replaced weekly

   c. Are linked to floc-well performance

   d. Distribute the flow over the clarifier surface

8. This modification to secondary clarifiers provides a zone of gentle mixing to improve sludge settling.
   a. Perimeter channel
   b. Flocculating center feed well
   c. Inlet channel mixers
   d. Riser pipes

9. A Stamford baffle
   a. Is typically made of concrete
   b. Attaches to the scum baffle
   c. Redirects currents downward
   d. May increase effluent TSS

10. Algae growth on secondary clarifier weirs should be controlled because
    a. Excessive growth can result in unbalanced flows.
    b. The roots can eat into the metal v-notches of the weir plate.
    c. Algae absorb vital nutrients from the process.
    d. Its messy appearance offends visitors to the facility.

11. An operator wants to increase the sludge withdrawal rate from an organ-pipe style clarifier with submerged telescoping valves. He or she should
    a. Decrease the sludge pumping rate
    b. Lower the water level in the sludge collection box
    c. Manually adjust each riser pipe telescoping weir
    d. Shift influent flow to a different secondary clarifier

12. With organ-pipe-style clarifiers, stacking the pipes vertically
    a. Increases access to the tops of the riser pipes for sampling
    b. Interferes with placement of Stamford baffles
    c. Limits the adjustment of the telescoping weirs
    d. Increases clarifier mixing over horizontally stacked pipes

13. Which of the following statements about Tow-Bro type sludge collection mechanisms is false?
    a. Openings along the manifold are largest at the clarifier edge.
    b. May use either a sludge wet well or pump directly from clarifier.
    c. Removes sludge from the entire clarifier floor at the same rate.
    d. Requires routine adjustment of the telescoping weirs.

14. In a plug flow activated sludge basin with tapered aeration, where would you expect to find the most diffusers?
    a. At the influent end of the basin
    b. Near the middle of the basin
    c. At the effluent end of the basin
    d. Equally spaced along the basin length

15. Nonporous aeration systems are
    a. More efficient than fine-bubble
    b. Have larger openings than fine-bubble
    c. Most widely used type of diffused air
    d. Consist of a header with nozzles

16. Jet aeration systems are classified as
    a. Fine bubble
    b. Coarse bubble
    c. Surface aeration
    d. Draft tubes

17. The nesting cones of a jet aeration header
    a. Inject air through the inner cone
    b. Create turbulence for oxygen transfer
    c. Require both air and MLSS flow at all times
    d. Typically point up toward the basin surface

18. The vertical drop pipes that connect the blowers to the diffusers at the bottom of the activated sludge basins are called
    a. Laterals
    b. Diffusers
    c. Draft tubes
    d. Downcomers

19. Surface aerators and rotors transfer oxygen into the MLSS by
    a. Injecting air through nozzles
    b. Agitating the water surface
    c. Conveying air supplied by a blower
    d. Recirculating MLSS over diffusers

20. Return activated sludge pumps are often equipped with _____ to allow the operator to adjust the flowrate.
    a. Upstream wet wells
    b. Magnetic flow meters
    c. Variable frequency drives
    d. Diaphragms

Process Variables for the Activated Sludge Basin

1. Return activated sludge is higher in concentration than the MLSS in the activated sludge basins.
    ☐ True
    ☐ False

2. Slow-settling sludges may continue to settle and compact in the settleometer test after 30 minutes.
    ☐ True
    ☐ False

3. The RAS is a fixed percentage of the influent flow. The clarifier blanket depth is less than 0.6 m (2 ft). The influent flow to the WRRF is increasing. What is happening to the clarifier SLR?
    a. Held constant by constant RAS percentage
    b. Decreasing with increasing influent flow
    c. Held constant by blanket depth
    d. Increasing with increasing influent flow

4. Increasing the RAS flowrate will
    a. Decrease the RAS concentration
    b. Decrease the SLR to the secondary clarifier
    c. Increase the WAS concentration
    d. Increase sludge age

5. The RAS pumping rate was increased and the concentration of MLSS in the activated sludge basin increased. What must be true?
    a. Before the increase, MLSS was settling in the activated sludge basins.
    b. After the increase, the SLR to the clarifier was too high.
    c. Before the increase, much of the MLSS was in the clarifier blanket.
    d. After the increase, increased exposure to BOD_5 increased the MLSS.

6. Waste activated sludge influences all of the following EXCEPT
    a. Sludge age
    b. Sludge settleability
    c. Sludge concentration
    d. Sludge blanket depth

7. An activated sludge system does not currently remove ammonia. To retain nitrifying bacteria and increase their numbers, which of the following actions could be taken?
    a. Increase return activated sludge
    b. Decrease sludge wasting
    c. Increase HDT
    d. Decrease clarifier SLR

8. The influent flow to the activated sludge basin is 8.7 ML/d (2.3 mgd). The RAS flow is 63% of the influent flow at 5.48 ML/d (1.45 mgd). If the MLSS concentration is 2600 mg/L, what is the expected RAS concentration?
    a. 4000 mg/L
    b. 4200 mg/L
    c. 6700 mg/L
    d. 7200 mg/L

9. After settling for 30 minutes in the settleometer, the SSV is 280 mL. What is the SSC if the starting MLSS concentration was 1850 mg/L?
    a. 5180 mg/L
    b. 6607 mg/L
    c. 7400 mg/L
    d. 9250 mg/L

10. The RAS concentration was calculated using flow data and the MLSS concentration. The calculation predicts a RAS concentration of 9000 mg/L. Filamentous bacteria are preventing the MLSS from settling and compacting well in the settleometer test. After 30 minutes, the SSC is only 6200 mg/L. What is the maximum RAS concentration achievable in the clarifiers if the sludge is allowed to remain in the clarifier for 30 minutes?
    a. 3100 mg/L
    b. 6200 mg/L
    c. 7600 mg/L
    d. 9000 mg/L

11. When solids are removed from the clarifier faster than they are added to the clarifier
    a. Blanket depth will increase
    b. Most of the MLSS will be in the clarifier
    c. Blanket depth will decrease
    d. Treatment efficiency will decrease

12. The RAS flow is currently set at 50% of influent flow. If the RAS flow is decreased to 30%, the operator should expect to see
    a. Reduced overall treatment time
    b. Rat holing of the clarifier sludge blanket
    c. Decreased solids loading to the clarifier
    d. Increase in RAS concentration

13. Mixed liquor suspended solids with a concentration of 1800 mg/L was placed into a settleometer and allowed to settle for 30 minutes. The initial sludge volume was 2000 mL and the 30-minute sludge volume was 280 mL. Find the SSC.
    a. 6429 mg/L
    b. 7143 mg/L
    c. 12 857 mg/L
    d. 14 286 mg/L

14. The return activated sludge flow is set at 70% of influent flow. If the MLSS concentration is 2600 mg/L, what is the anticipated RAS concentration assuming solids are not being stored in the clarifier blanket?
    a. 6300 mg/L
    b. 3700 mg/L
    c. 7500 mg/L
    d. 9000 mg/L

15. The operator knows from the settleometer test that the SSC at 30 minutes is 10 000 mg/L. The initial MLSS concentration was 2500 mg/L. If the operator wants to achieve the thickest RAS concentration for this sludge, but does not want to leave the RAS in the clarifier any longer than required, what should the RAS flowrate be when the influent flow is 4200 m³/d (1.1 mgd)?
    a. 1400 m³/d (0.37 mgd)
    b. 2500 m³/d (0.66 mgd)
    c. 2800 m³/d (0.74 mgd)
    d. 5100 m³/d (1.35 mgd)

Sludge Age

1. In simplest terms, sludge age is simply the length of time it takes to drain an activated sludge basin.
    ☐ True
    ☐ False

2. Because yield, the mass of MLVSS produced per mass of BOD, is nearly constant for domestic wastewater, the MLSS concentration will increase and decrease with the influent BOD load even when sludge age is held constant.
    ☐ True
    ☐ False

3. This method is recommended for activated sludge process control.
    a. Sludge age
    b. Constant MLSS mass
    c. Constant F/M
    d. RAS control

4. Gould sludge age was developed in the 1920s and was based on the idea that
    a. Activated sludge is primarily a physical–chemical treatment process.
    b. Activated sludge is primarily a biological treatment process.
    c. Influent soluble BOD is critical to activated sludge performance.
    d. Influent TSS is transformed in the activated sludge basin to MLSS.

5. Gould sludge age is calculated by dividing the mass of MLSS in the activated sludge basins by
    a. Mass of WAS leaving the system
    b. Mass of influent TSS
    c. Mass of influent BOD
    d. Mass of RAS in the return

6. In the term *MCRT*, mean is another word for
    a. Aggressive
    b. Average
    c. Acclimated
    d. Aerodynamic

7. Controlling the amount of WAS leaving the system each day controls the amount of MLVSS available to treat the incoming organic load. Therefore, controlling sludge age also controls
    a. Growth rate
    b. Yield
    c. Influent TSS
    d. Return rates

8. Sludge age is held constant. If the influent BOD concentration increases, the MLSS concentration will
    a. Remain about the same
    b. Increase proportionally to the load
    c. Decrease in percent volatile solids
    d. Equal the influent BOD time sludge age

9. The desired MCRT is 10 days. If MLSS is removed directly from the activated sludge basin, what percentage of the MLSS must be removed each day?
    a. 5%
    b. 10%
    c. 20%
    d. 25%

10. Under what condition will maintaining a constant MLSS concentration approximate constant sludge age?
    a. Volume of WAS removed each day is constant.
    b. RAS is a fixed percentage of influent flow.
    c. Effluent solids remain below 30 mg/L.
    d. Influent loads are nearly constant.

11. One advantage of using MCRT over SRT is
    a. Additional samples are required
    b. 100% of the sludge inventory is accounted for
    c. Difficulty in sampling clarifier sludge blankets
    d. Enables sludge storage in the clarifiers

12. Estimate the desired wasting rate from the figure below given the following information: MLSS concentration is 2000 mg/L, target sludge age is 7.5 days, WAS concentration is 6250 mg/L, and the activated sludge basin holds 2000 m³ (0.53 mgd).
    a. 50 m³ (0.013 mgd)
    b. 160 m³/d (0.044 mgd)
    c. 320 m³/d (0.088 mgd)
    d. 88 m³/d (0.023 mgd)

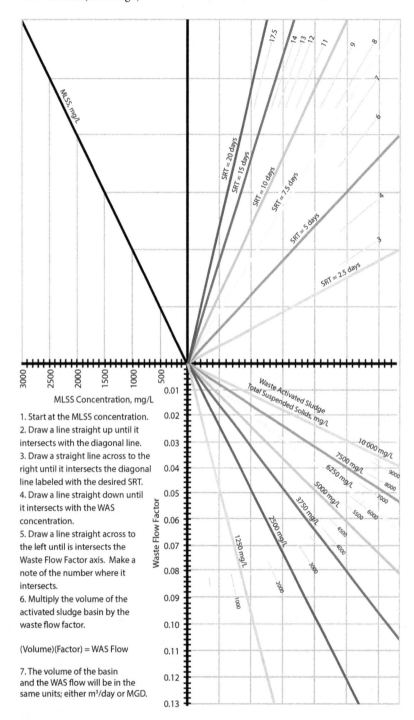

Graphical Tool for Estimating Wasting Rate for a Given Target SRT (U.S. EPA, 1977)

13. The SRT is currently at 6 days. The target SRT is 8 days. The operator should
    a. Increase the WAS
    b. Decrease the WAS
    c. Increase the RAS
    d. Decrease the RAS

14. Mean cell residence time and SRT are both calculated by dividing the mass of sludge in the system by the mass of sludge _____.
    a. In the influent
    b. In the effluent
    c. Leaving the system
    d. In the digester

15. An activated sludge process contains 8167 kg (18 000 lb) of MLSS in the activated sludge basins and another 2269 kg (5000 lb) in the secondary clarifiers. If the operator removes 907 kg (2000 lb) of MLSS in the WAS each day, what is the MCRT?
    a. 3.6 days
    b. 4 days
    c. 9 days
    d. 11.5 days

16. Given the following information, find the SRT.

    Activated sludge basin volume: 11 355 m³ (3 mil. gal)
    MLSS concentration: 2500 mg/L
    Wasting rate: 436 m³/d (80 gpm)
    WAS/RAS concentration: 8000 mg/L
    a. 5 days
    b. 8 days
    c. 11 days
    d. 14 days

Potential Effects of Foaming Events on Sludge Age

1. Wasting should be done all at once at the end of each week rather than daily to minimize the number of visits to the treatment facility.
    ☐ True
    ☐ False

2. Solid retention time calculations and wasting rate adjustments (if needed) should be done
    a. Daily
    b. Weekly
    c. Every 2 weeks
    d. Monthly

3. On Monday, Tuesday, and Wednesday, the operator wasted 1000 kg (2205 lb) from the system each day. The current SRT is 12 days, but the target sludge age is 8 days. What is the maximum amount of sludge the operator can waste on Thursday while staying within the 15% rule?
    a. 880 kg (1874 lb)
    b. 1150 kg (2536 lb)
    c. 5000 kg (11 025 lb)
    d. 15 000 kg (33 069 lb)

4. An activated sludge facility uses large quantities of ferric chloride in a downstream process to precipitate phosphorus. The precipitated phosphorus and iron, now iron sludge, is removed with tertiary filters. When the filters are cleaned, iron sludge is recycled back to the activated sludge process. How might the operator compensate for the changes in total MLSS concentration caused by these nonbiological solids?

    a. Increase blower output to increase available oxygen
    b. Decrease sludge age to remove solids faster
    c. Switch to constant mass for process control
    d. Substitute MLVSS for MLSS in the SRT calculation

5. All of the following may affect the accuracy of an SRT calculation EXCEPT

    a. Accuracy of WAS flow meter
    b. Lack of a blanket in the clarifier
    c. Foaming in the activated sludge basin
    d. High-effluent suspended solids

6. Solids in the clarifier blanket should be included in the sludge age calculation when

    a. The MLSS is settling and compacting well.
    b. During peak wet weather events.
    c. Clarifier sludge blankets are deep.
    d. Clarifiers are small compared to the basins.

7. Before collecting grab samples from the final effluent for process control calculations, operators should

    a. Check with their regulatory agency
    b. Verify that the effluent is clean
    c. Reduce RAS and WAS flows
    d. Remove algae from clarifier weirs

8. An activated sludge process has extremely large basins with a combined HDT of 28 hours. The clarifiers are relatively small and have an HDT of only 2 hours. The operator maintains the blankets below 3 ft of depth. Which of the following calculations for sludge age is most appropriate for this facility?

    a. GSA
    b. MCRT
    c. SRT
    d. SRT$_{aerobic}$

9. A conventional activated sludge process has an HRT of 6 hours. There are three clarifiers online with a combined detention time of 4 hours. Which of the following sludge age calculations is most appropriate for this system?

    a. GSA
    b. MCRT
    c. SRT
    d. SRT$_{aerobic}$

Food-to-Microorganism Ratio

1. Setting a target SRT or MCRT also determines the F/M.
    ☐ True
    ☐ False

2. The activated sludge basin volume is 30 000 m³ (7.9 mil. gal). If the MLSS concentration is 2200 mg/L and is 80% volatile, how many kilograms (pounds) of MLVSS is in the basin?

    a. 52 800 kg (115 959 lb)
    b. 66 000 kg (144 949 lb)
    c. 82 500 kg (181 867 lb)
    d. 97 000 kg (231 788 lb)

3. A WRRF receives 817 kg (1800 lb) of BOD_5. If the primary clarifier removes 35% of the incoming BOD_5, what is the load to the activated sludge process?

    a. 286 kg (630 lb)

    b. 531 kg (1170 lb)

    c. 1256 kg (2769 lb)

    d. 2234 kg (5142 lb)

4. A conventional activated sludge system that performs only BOD_5 removal typically has an F/M in the range of

    a. 0.05 to 0.15

    b. 0.2 to 0.4

    c. 0.4 to 0.6

    d. 0.8 to 1.2

5. Disadvantages of using F/M for process control include

    a. Helps maintain constant microorganism growth rate

    b. Requires influent ammonia monitoring for nitrfying facilities

    c. Provides a clear mechanism for determining wasting

    d. Forces operator to monitor and predict the influent load

Relationships Between Process Variables

1. Sludge age and F/M can be controlled independently.

    ☐ True

    ☐ False

2. Operators may increase the total mass of live microorganisms in their process by increasing sludge age.

    ☐ True

    ☐ False

3. The SRT is currently 14 days. The operator increases the wasting rate to lower the SRT to 10 days. Which of the following variables will decrease?

    a. F/M

    b. Percentage of MLVSS

    c. MLSS concentration

    d. BOD load

4. Mixed liquor suspended solids concentrations increase with increasing SRT, but the percentage of MLVSS decreases because

    a. Live microorganisms are wasted faster than inert material.

    b. Inert material accumulates in the process unchanged.

    c. Floc sizes increase at older SRTs and keep grit suspended.

    d. More BOD_5 per microorganism.

5. The influent BOD load is unchanged. Solids retention time is increasing. What must be true?

    a. Growth rate is decreasing.

    b. Yield is increasing.

    c. MLSS concentration is decreasing.

    d. Wasting rate is increasing.

6. The SRT is currently 12 days. If the SRT is decreased to 8 days, what will happen to the F/M?

    a. It will increase by 50%.

    b. It will remain the same.

    c. It will increase.

    d. It will decrease.

7. Maintaining a constant MLSS concentration as a process control strategy approximates constant SRT when
   a. F/M is between 0.2 and 0.4 kg BOD_5/kg MLVSS (lb/lb).
   b. The WRRF is below 50% of its rated capacity.
   c. MLSS does not change more than 200 mg/L per day.
   d. Influent BOD_5 and TSS loads are nearly constant.

8. This is almost never measured for WRRF influent, but it can have a big effect on the MLSS concentration.
   a. BOD_5
   b. TSS
   c. Ammonia
   d. VSS

9. Match the process control strategy to its primary disadvantage over other methods.
   a. MCRT
   b. SRT
   c. $SRT_{aerobic}$
   d. F/M
   e. Constant MLSS

   1. Operator must predict influent loads
   2. Requires more data collection
   3. Assumes no solids in the clarifier blanket
   4. If influent loads vary, SRT and F/M will also vary
   5. Growth of filamentous bacteria tied to total SRT

Process Control for Activated Sludge Basins

1. An effluent soluble BOD_5 concentration below 10 mg/L is an indication of well-flocculated MLSS.
   ☐ True
   ☐ False

2. Sludge bulking can result when the sludge age is either too high or too low.
   ☐ True
   ☐ False

3. Conventional and extended aeration are the two most common operating modes for domestic facilities. Check the box to indicate which of these operating modes has a higher HDT, SRT, F/M, or volumetric loading rate.

	Conventional	Extended Air
Detention time, hours		
SRT, days		
F/M kg BOD_5/kg MLVSS		
Volumetric loading rate		

4. The minimum sludge age needed to produce well-flocculated MLSS is around
   a. 2 to 3 days
   b. 5 to 15 days
   c. 15 to 30 days
   d. Greater than 30 days

5. Large, dense, well-flocculated MLSS should
   a. Produce low effluent ammonia concentrations
   b. Contain an abundance of filaments
   c. Remove particulates from the water as they settle
   d. Leave a cloudy supernatant behind

6. White foam is associated with
   a. Poorly flocculating activated sludge
   b. *Beggiatoa* filaments
   c. SRTs lower than 2 days
   d. F/M greater than 0.4

7. For filamentous bacteria to proliferate in activated sludge, environmental conditions must support their growth and
   a. Sludge age must be long enough
   b. DO must be low
   c. Volatile fatty acids are present
   d. Grease concentrations are high

8. How fast a floc particle settles in the secondary clarifier is influenced by all of the following EXCEPT
   a. Water temperature
   b. Presence of filaments
   c. Solids concentration
   d. DO concentration

9. Some facilities track their SSVs after 5 minutes because this is a good measure of sludge settling velocity. Why do many facilities see the SSV_5 decrease in summer and increase in winter?
   a. Loading changes with the seasons
   b. Water temperature affects settling velocity
   c. Typically more filaments in the summer
   d. Higher MLSS concentrations in summer

10. Which MLSS will tend to settle the slowest?
    a. MLSS of 3000 mg/L, few filaments present
    b. MLSS of 5000 mg/L, many filaments present
    c. MLSS of 3000 mg/L, many filaments present
    d. MLSS of 5000 mg/L, few filaments present

11. If MLSS has an SVI greater than 200 mL/g, it is defined as
    a. Well flocculated
    b. Filament free
    c. Well compacted
    d. Bulking

12. The maximum MLSS concentration for most activated sludge systems is limited by
    a. Bioflocculation
    b. Basin size
    c. Clarifier capacity
    d. Predation

13. The washout SRT for a newly discovered microorganism is 8 days at 20 °C. To keep this organism in the system, the target SRT should be at least
    a. 8 days
    b. 12 days
    c. 16 days
    d. 28 days

14. To prevent the growth of the filament *Sphaerotilus natans* in activated sludge, low DO conditions must be avoided and the SRT should be kept below about (assume a water temperature of 20 °C [68 °F])
    a. 3 days
    b. 9 days
    c. 15 days
    d. 21 days

Setting a Target Dissolved Oxygen Concentration

1. When DO concentrations are low, certain types of filaments grow faster than the floc-forming bacteria because
   a. DO is toxic to filament formers
   b. Floc-former growth slows down
   c. Filaments are protected by floc formers
   d. Sludge age increases

2. If the F/M is 0.4 kg BOD$_5$/kg MLVSS, what DO concentration will prevent the growth of low DO filaments?
   a. 0.5 mg/L
   b. 1 mg/L
   c. 2 mg/L
   d. 3 mg/L

3. An activated sludge facility is not required to remove ammonia. The facility's F/M is only 0.2 kg BOD$_5$/kg MLVSS. What is the most likely outcome of maintaining a DO concentration of 1.0 mg/L in the activated sludge basins?
   a. Growth of low DO filaments
   b. Floc-forming bacteria dominate
   c. Sludge bulking
   d. Low effluent ammonia

4. An activated sludge process is experiencing extreme foaming and settleability is poor. Reviewing records, the chief facility operator discovers that the sludge age has increased over the last week to 16 days. Operators know from experience that the foaming subsides as long as the SRT is well controlled below 12 days. They make the necessary adjustments to bring the SRT below 12 days. How long will it take for the process to completely recover and stabilize at the new sludge age?
   a. Less than 12 days
   b. Between 24 and 36 days
   c. Between 36 and 50 days
   d. Greater than 75 days

5. Which of the following filaments grow in response to low DO conditions?
   a. *S. natans*
   b. Type 021N
   c. *M. parvicella*
   d. Nocardioforms

6. The SVI has increased over the last week from 120 to 180 mL/g. A microscopic evaluation reveals many *H. hydrossis* filaments. The operator should
   a. Decrease sludge age
   b. Decrease DO concentrations
   c. Increase sludge age
   d. Increase DO concentrations

Process Variables for Secondary Clarifiers

1. This process variable is typically what limits clarifier capacity.
   a. SOR
   b. Weir loading rate
   c. RAS flowrate
   d. SLR

2. Find the SOR for a clarifier that is 19.8 m (65 ft) in diameter when the influent flow is 7.57 ML/d (2 mgd) and the RAS is set at 60% of influent flow.

    a. 9.8 m³/m²·d (241 gpd/sq ft)

    b. 24.6 m³/m²·d (603 gpd/sq ft)

    c. 14.7 m³/m²·d (362 gpd/sq ft)

    d. 39.3 m³/m²·d (965 gpd/sq ft)

3. The RAS flow is increased from 50% of influent flow to 65% of influent flow. Treatment time in the activated sludge basins

    a. Decreases with increasing RAS flow

    b. Should only be adjusted in 10% increments

    c. Remains the same regardless of RAS flow

    d. Increases with increasing RAS flow

4. Find the SOR for a clarifier that is 36.6 m (120 ft) in diameter when the influent flow is 26.5 ML/d (7 mgd).

    a. 14.7 m³/m²·d (348 gpd/sq ft)

    b. 25.2 m³/m²·d (619 gpd/sq ft)

    c. 36.3 m³/m²·d (892 gpd/sq ft)

    d. 56.7 m³/m²·d (1396 gpd/sq ft)

5. A rectangular secondary clarifier has a total weir length of 18.3 m (60 ft). The influent flow is 5 ML/d (1.32 mgd). Find the weir overflow rate.

    a. 205 m³/m·d (16 500 gpd/ft)

    b. 234 m³/m·d (18 857 gpd/ft)

    c. 273 m³/m·d (22 000 gpd/ft)

    d. 328 m³/m·d (26 400 gpd/ft)

6. A secondary clarifier that is 16.8 m (55 ft) in diameter is equipped with an inset launder. The outermost weir is 0.75 m (2.5 ft) from the outer wall of the clarifier. What is the diameter of the outer weir?

    a. 15.3 m (50 ft)

    b. 16.1 m (52.5 ft)

    c. 16.8 m (55 ft)

    d. 18.3 m (60 ft)

7. A WRRF receives 11.4 ML/d (3 mgd) of influent flow. The MLSS concentration is 3000 mg/L and the RAS concentration is 7600 mg/L. Find the total mass going to the clarifier if the RAS is set at 65% of influent flow.

    a. 46 170 kg (101 331 lb)

    b. 56 430 kg (123 849 lb)

    c. 90 249 kg (198 909 lb)

    d. 143 245 kg (314 386 lb)

8. Find the SLR to a clarifier if the mass of MLSS going to the clarifier is 20 000 kg (44 080 lb) and the surface area is 117 m² (1260 sq ft).

    a. 53 kg/m²·d (10.8 lb/hr/sq ft)

    b. 92 kg/m²·d (18.8 lb/hr/sq ft)

    c. 117 kg/m²·d (23.9 lb/hr/sq ft)

    d. 171 kg/m²·d (35.0 lb/hr/sq ft)

9. If the SLR is too high

    a. Blanket depth will increase

    b. RAS thickness will increase

    c. Blanket depth will decrease

    d. RAS thickness will decrease

10. Last week, the SVI was 120 mL/g. This week, the SVI increased to 210 mL/g. The operator may need to
    a. Increase the RAS flowrate
    b. Stop adding polymer to the clarifier
    c. Remove a clarifier from service
    d. Reduce the SLR

11. A clarifier has a SLR of 171 kg/m²·d (35 lb/d/sq ft). The RAS concentration is 13 000 mg/L. Use the diagram in the figure below to determine which of the following statements is true.
    a. The clarifier can settle sludge with an SVI up to 150 mL/g.
    b. The underflow rate is between 12 and 16 m³/m²·d (300 and 400 gpd/sq ft).
    c. If the current, operating SVI is 85 mL/g, the clarifier will be overloaded.
    d. Because the operating point does not fall directly on a line, the chart cannot be used.

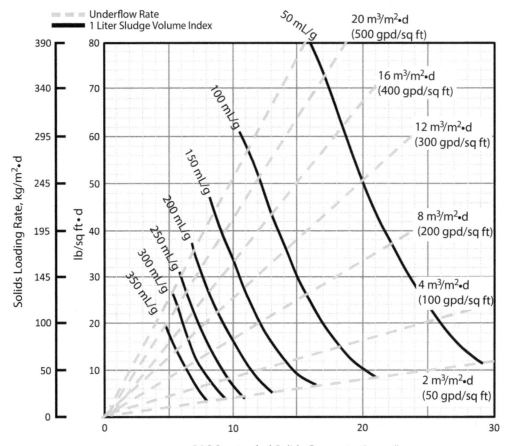

Modified from Daigger, 1995

Process Control for Secondary Clarifiers

1. Return activated sludge flowrates should be kept as low as possible while meeting treatment goals to
    a. Maximize treatment time in the activated sludge basin
    b. Build a deep sludge blanket to squeeze the water from the RAS
    c. Minimize potential for denitrification to take place
    d. Avoid sending excess water to solids handling processes

2. One advantage of flow pacing the RAS pumps to the WRRF influent flow is
    a. Variable diurnal blanket depth
    b. Nearly constant RAS concentration
    c. Maximizes RAS concentration at low flow
    d. Solids may be stored in the clarifier

3. Sludge settleability has deteriorated over the last 2 weeks and the clarifiers are building deep blankets. The operator turns up the RAS flowrate to remove the sludge from the clarifier faster, but instead of going down, the blankets get even deeper. This is because
    a. RAS pumps were not turned up enough
    b. RAS pumps need time to catch up, then the blanket will drop
    c. Increasing RAS made the overloading problem worse
    d. SOR is too low

4. The secondary clarifier at a small facility is overloaded. There are no other clarifiers available. A short-term solution may be to
    a. Adjust SVI with chemicals
    b. Increase wasting to reduce SRT
    c. Build additional clarifiers
    d. Increase RAS flow

5. Gas bubbles generated in the sludge blanket may cause
    a. Short-circuiting
    b. Ashing or clumping
    c. Air binding
    d. Ratholing

6. A key indicator that denitrification is taking place in a circular secondary clarifier blanket is
    a. RAS concentration decreases with increasing pump speed
    b. Flocculation center well is full of heavy, thick foam
    c. Sludge collecting between scum baffle and weir
    d. Scum on clarifier surface turns from brown to green

Operation—Daily Operation

1. An activated sludge process may be started up using seed sludge from a nearby anaerobic digester.
    ☐ True
    ☐ False

2. Water temperature in the activated sludge basins should be measured daily in facilities that are required to remove ammonia.
    ☐ True
    ☐ False

3. An operator has just finished the daily walkthrough of the activated sludge process. He or she decides to increase the RAS flow slightly. What did the operator observe that resulted in this decision?
    a. MLSS not compacting well in the settleometer test
    b. Clarifier blankets deeper than 0.7 m (2 ft)
    c. Sludge age is much higher than the target
    d. RAS concentration is too low

4. During startup, when seed sludge is not available, influent wastewater is preferred over primary effluent because
    a. Higher BOD concentrations in influent
    b. Particulate matter is needed for floc formation
    c. Primary clarifier not yet placed into service
    d. Greater populations of microorganisms

5. Mixed liquor suspended solids should not be wasted from the process for the first time until
   a. Minimum startup concentration has been reached
   b. MLSS concentration begins to overload the clarifier
   c. Supernatant BOD and TSS are below 10 mg/L
   d. Rotifers and stalked ciliates are observed

6. A primary clarifier with four groundwater pressure relief valves is taken out of service. If two of the valves fail to open with high groundwater, what is the likely outcome?
   a. Sump pump may be needed to remove groundwater
   b. Clarifier may move or be pushed upward
   c. Clarifier may overflow as water backs up into it
   d. Operator will not be able to reach backflush

Data Collection, Sampling, and Analysis

1. An operator notices that one area of the activated sludge basin is particularly turbulent. The most likely cause is
   a. Damaged or missing diffuser
   b. Clogged air header
   c. Blower output is too high
   d. Water level is too low

2. An activated sludge basin is covered with large amounts of crisp, white foam. This indicates
   a. Presence of *Nocardia*
   b. Startup conditions
   c. Low organic loading rate
   d. Nutrient deficiency

3. When collecting a clarifier core sample, the sample is collected when the rake arm is at a 90-deg angle to the sampling location for this reason.
   a. Avoids tangling the sludge judge and clarifier mechanism
   b. Allows the sludge blanket to resettle after being disturbed by the rake arm
   c. Ensures that the lowest possible blanket depth is sampled
   d. Prevents the sludge blanket from getting too thick to sample

4. An activated sludge process has three identical secondary clarifiers. Sludge blanket depths in each are 1 m, 1.2 m, and 3 m. What is the most likely cause for the differences?
   a. Sludge is being removed faster from the third clarifier.
   b. The RAS pump on the first clarifier has failed.
   c. Uneven flow distribution feeding the clarifiers.
   d. WAS is only being removed from the second clarifier.

5. In the settleometer test, the SSV at 30 minutes is an indicator of
   a. Sludge settling velocity
   b. Sludge compaction
   c. Sludge concentration
   d. Sludge toxicity

6. This type of container should be used for the settleometer test.
   a. Graduated cylinder
   b. 1-L beaker
   c. 5-gal bucket
   d. Mallory settleometer

7. When collecting samples for the settleometer test
    a. Shake vigorously to ensure they are well mixed
    b. Minimize shaking and agitation
    c. Place samples on ice and cool to 4 °C
    d. Add 2 mL of nitric acid as a preservative

8. Settled sludge may rise back to the top of the settleometer test for this reason.
    a. Breakdown of solids and production of methane gas
    b. Warming during the test releases dissolved gases
    c. Denitrification and production of nitrogen gas
    d. Hydraulic currents in the settleometer

9. Two settleometer tests are run side by side. The first settleometer contains undiluted MLSS at 4000 mg/L. The second settleometer contains 50% MLSS and 50% secondary clarifier effluent for a concentration of 2000 mg/L. Both settleometer tests are started at the same time. After 5 minutes, the SSV in the first settleometer is 900 mL but 750 mL in the second. What must be true?
    a. Filaments are causing poor settleability in the first test.
    b. Hindered settling is taking place in the first settleometer.
    c. Wasting should be decreased.
    d. The RAS should be increased.

10. The 30-minute settled volume is 320 mL and the MLSS concentration is 2950 mg/L. Find the SVI in milliliters per gram.
    a. 108 mL/g
    b. 222 mL/g
    c. 320 mL/g
    d. 512 mL/g

11. The MLSS concentration at the start of the settleometer test was 2500 mg/L. After settling for 30 minutes, the SSV is 230 mL. What is the settled sludge concentration?
    a. 2500 mg/L
    b. 7813 mg/L
    c. 10 870 mg/L
    d. 22 609 mL/g

12. The MLSS concentration in the activated sludge basin is 3200 mg/L. The influent flow is 2 mgd and the RAS is flow paced at 60% of the influent flow. What is the RAS concentration?
    a. 4160 mg/L
    b. 5120 mg/L
    c. 3870 mg/L
    d. 8530 mg/L

13. The 5-minute SSV is dependent on
    a. Water temperature
    b. Presence of filaments
    c. MLSS concentration
    d. All of the above

14. The primary difference between an OUR and a SOUR test is
    a. OUR is conducted at 25 °C and SOUR at 22 °C.
    b. SOUR corrects for solids concentration.
    c. OUR testing is used with anaerobic digesters.
    d. SOUR is a shorter test.

15. The SOUR test measures
    a. Blower and diffuser efficiency
    b. DO probe fouling
    c. How fast microorganisms are consuming oxygen
    d. Effluent soluble BOD_5

16. Mixed liquor volatile suspended solids are used for OUR and SOUR testing rather than MLSS because
    a. The MLSS analysis takes longer.
    b. The MLVSS test is more accurate.
    c. The MLVSS test measures live microorganisms.
    d. The MLSS test does not include inert solids.

17. The SOUR for a particular WRRF is typically between 5 and 8 mg O_2/L·h. When the operator tested it at 3:00 p.m. today, the SOUR had increased to 12 mg O_2/L·h. What is the most likely cause for the increase?
    a. Toxic or inhibitory substance entered the activated sludge basin
    b. SRT dropped below the minimum required to support nitrification
    c. Increased BOD_5 loading to the activated sludge basin
    d. Number of filamentous bacteria doubled

18. A clarifier is operating with a SOR of 8 m³/m²·d (200 gpd/sq ft) and the RAS concentration is 9000 mg/L. Using the chart in the figure below, find the maximum theoretical SVI the MLSS can have without overloading the clarifier.
    a. 150 mL/g
    b. 200 mL/g
    c. 250 mL/g
    d. 300 mL/g

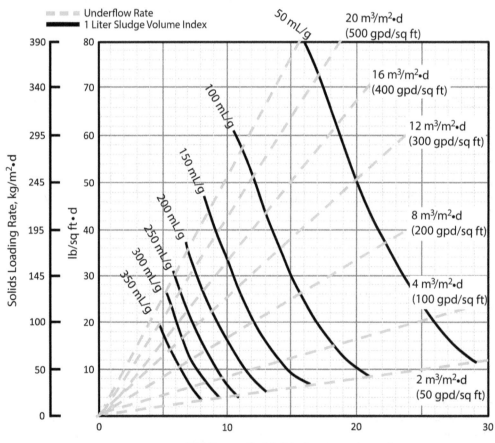

Modified from Daigger, 1995

Maintenance

1. Process control consists of setting goals and defining allowable tolerances and deviations from those goals.
   - ☐ True
   - ☐ False

2. For best results when troubleshooting, operators should take multiple corrective actions at the same time.
   - ☐ True
   - ☐ False

3. After lunch, an operator is conducting a walkthrough of their facility and they notice billowing clouds of sludge at the edges of the secondary clarifiers. What is the most likely cause?
   - a. Solids overloading
   - b. Hydraulic surge
   - c. Denitrification
   - d. RAS rate too low

4. A facility has one clarifier that is 16.8 m (55 ft) in diameter and another that is 22.9 (75 ft) in diameter. How should the flow be split so that each clarifier has the same hydaulic and SLR?
   - a. 35% to the smaller clarifier
   - b. 54% to the larger clarifier
   - c. 73% to the smaller clarifier
   - d. 73% to the larger clarifier

5. Classify the action taken by whether it is process control or troubleshooting.
   - a. SRT target set at 10 days
   - b. Microscopic examination done to determine cause of foaming
   - c. RAS flowrate adjusted to keep clarifier blanket below 3 ft
   - d. Blower output increased to maintain DO of 2 mg/L
   - e. Second clarifier placed into service to reduce SOR
   - f. Adding nutrients to correct sludge bulking

Aeration System Troubleshooting

1. The surface of an aerated activated sludge basin is smooth at one end, while the other end has large bubbles coming up in one corner. What is the most likely cause?
   - a. Air rates too high at one end of the basin
   - b. Broken or missing diffuser in corner of basin
   - c. Valves need adjusting to balance air distribution
   - d. Diffusers are not at the same elevation

2. To remove a blockage from a jet aeration system nozzle
   - a. Backflush by filling the header with air
   - b. Increase pump discharge pressure
   - c. Recirculate MLSS at increased rate
   - d. Insert a wire coat hanger and pull

3. An activated sludge basin equipped with mechanical aerators has DO concentrations of 3 mg/L near the aerators, but only 0.25 mg/L at the edges of the basin. The operator should
   - a. Decrease aerator speed
   - b. Lower the tank weir setting
   - c. Decrease aerator submergence
   - d. Increase aerator speed

4. Flooding occurs in a mechanical aerator equipped with a draft tube when
   a. Tank weir is too low
   b. Draft tube flow direction is reversed
   c. Aerator is operating at high speed
   d. Draft tube detached from aerator

5. The DO concentration in the activated sludge basin increases from 2 to 6 mg/L over a 2-hour period. What is the most likely cause?
   a. Increased BOD loading
   b. Toxic material in influent
   c. Nitrification occuring
   d. Aerator flooding

Sludge Quality Problems

1. Match the foam color to its most likely cause.

   1. Thin, white-to-gray foam   a. Old sludge
   2. Billowing, white foam   b. Low F/M
   3. Greasy, dark tan foam   c. Nutrient deficiency
   4. Thick, greasy, chocolate-brown foam with crust   d. Septic conditions
   5. Dark or black foam   e. Type 1863
   6. Sticky, viscous foam   f. Recycle of fines
   7. Volcanic or pumice-like foam   g. Nocardioforms
   8. Thick, scummy dark-brown foam   h. *M. parvicella*

2. When starting up a new facility, how long does it typically take for startup foam to go away?
   a. About 24 hours
   b. 2 to 4 days
   c. 5 to 7 days
   d. 2 weeks

3. A storm event caused influent flow to the WRRF to triple. After the event, white foam is observed on the surface of the activated sludge basins and the RAS concentration has decreased. Why?
   a. Solids loss from clarifier lowered SRT
   b. Toxic compound in stormwater flow
   c. RAS return flowrate too high
   d. MLSS pH dropped below 6.5

4. Brown foam should be removed from activated sludge basin and clarifier surfaces to
   a. Improve overall facility appearance
   b. Prevent flooding of mechanical aerators
   c. Keep the wind from blowing it onto walkways
   d. Reduce numbers of filamentous bacteria

5. An activated sludge system is required to remove ammonia. The water temperature is currently 15 °C and the $SRT_{aerobic}$ is 24 days. Thick, greasy, chocolate brown foam covers the surface of the activated sludge basins and has completely filled the clarifier floc wells. The operator should
   a. Chlorinate the RAS
   b. Manually remove foam and reduce SRT
   c. Decrease DO concentrations in the basins
   d. Increase clarifier rake arm speed

6. The MLSS gradually turned from light tan to almost black over several hours. Dissolved oxygen concentrations are normal and excess foaming is not present. No sidestreams were entering the basin during this time. What could have caused the color change?
   a. Septic conditions
   b. Sludge recycle from digester

    c. Powdered activated carbon

    d. Type 021N filament present

7. The surface of the activated sludge basin and clarifier are covered with brown foam. Microscopic examination shows there are almost no filaments in the foam. What must be true?

    a. *M. parvicella* is causing the foaming.

    b. The SRT is long and the F/M is low.

    c. Nocardioforms are causing the foaming.

    d. Denitrification is occurring in the clarifier.

8. This foam-forming filamentous bacteria tends to cause problems during colder weather.

    a. Nocardioform

    b. Type 021N

    c. Type 1863

    d. *M. parvicella*

9. A WRRF is required to remove ammonia. The activated sludge basins are covered with thick, greasy, brown foam. The water temperature in the basin is currently 10 °C. How low can the operator drop the aerobic SRT before they risk losing nitrification?

    a. 4 days

    b. 6 days

    c. 9 days

    d. 15 days

10. A domestic WRRF receives a majority of its wastewater from a fruit processing and canning factory. The WRRF has an ongoing issue with slimy, viscous, gray foam. To fix the problem, they need to

    a. Remove foam with a vac-truck

    b. Increase DO concentrations in the basin

    c. Add missing nutrients

    d. Add polymer to help settle fines

11. Excessive fines from septage receiving or solids handling processes generate a pumice-like foam because

    a. Fines don't flocculate well

    b. Promotes the growth of Type 1947

    c. Polymer in recycle causes foaming

    d. Vital nutrients are missing

Secondary Clarifiers

1. Sludge with an SVI of 225 mL/g would be considered bulking sludge.

    ☐ True

    ☐ False

2. This condition results when the MLSS fails to flocculate and leaves behind a turbid supernatant.

    a. Zoogleal formations

    b. Dispersed growth

    c. Pin floc

    d. Straggler floc

3. A sample is collected from the end of an activated sludge process for a settleometer test. After 30 minutes, the supernatant is crystal clear, but the sludge has only settled to a volume of 500 mL. What is the most likely cause?

    a. Pin floc

    b. High DO shear

    c. Filaments

    d. Dispersed growth

4. The first step to solving a filament problem is to
    a. Increase the wasting rate
    b. Raise DO concentrations to 3 mg/L
    c. Adjust sludge age
    d. Identify the filament by staining

5. Which two chemicals may be added to activated sludge for filament control?
    a. Ferric chloride and PACl
    b. Chlorine and PACl
    c. PACl and sodium chlorate
    d. Chlorine and ferric chloride

6. This term is used to describe the total amount of a chemical, such as chlorine, to the treatment process.
    a. Dose
    b. Demand
    c. Residual
    d. Ion

7. When should chlorine addition to an activated sludge process be stopped?
    a. After 3 days of dosing
    b. When only empty sheaths remain
    c. SVI reaches 225 mL/g
    d. MLSS has deflocculated

8. Which of the following chemicals is most effective for controlling *M. parvicella*?
    a. Ferric chloride
    b. Chlorine
    c. Polyaluminum chloride
    d. Sodium chlorate

9. Chlorine is often applied directly to nocardioform foam rather than in the RAS line because
    a. Filaments concentrate in the foam
    b. Easier to ensure complete coverage
    c. Few facilities have ports on their RAS lines
    d. RAS exerts too much demand

10. If the MLSS concentration gets too low, this condition may result.
    a. Pinpoint floc
    b. Straggler floc
    c. Rapid settling
    d. Slow settling

11. If a secondary clarifier is already overloaded with solids, turning up the RAS flowrate will
    a. Reduce the blanket depth
    b. Decrease the SLR
    c. Increase the blanket depth
    d. Decrease overall treatment time

References

Daigger, G. T. (1995) Development of Refined Clarifier Operating Diagrams Using an Updated Settling Characteristics Database. *Water Enviroment Research, 67,* 95.

U.S. Environmental Protection Agency. (1977). *Process Control Manual for Aerobic Biological Wastewater Treatment Facilities.* U.S. Environmental Protection Agency: Washington, D.C.

CHAPTER 9
Nutrient Removal

Nitrogen and Phosphorus in Wastewater

1. Which of the following nitrogen compounds is not typically found in influent wastewater?
    a. Ammonia
    b. Nitrite
    c. Organic nitrogen
    d. TKN

2. Nitrate is sometimes added to the collection system to
    a. Control odors
    b. Precipitate phosphorus
    c. Remove BOD
    d. Increase biomass

3. The wastewater from a WRRF contains 2 mg/L of organic nitrogen, 5 mg/L of nitrate-nitrogen, and 1 mg/L of ammonia-nitrogen. How much TKN is leaving the facility?
    a. 1 mg/L as N
    b. 2 mg/L as N
    c. 3 mg/L as N
    d. 7 mg/L as N

4. After biological treatment is complete, most of the remaining phosphorus will be present as
    a. Orthophosphate
    b. Polyphosphate
    c. Organic phosphate
    d. Biomass

5. The influent to a WRRF contains 300 mg/L BOD_5 and 7 mg/L TP. If neither chemical nor EBPR are used, what is the expected effluent P concentration?
    a. 1 mg/L as P
    b. 2 mg/L as P
    c. 4 mg/L as P
    d. 7 mg/L as P

6. A conventional activated sludge treatment process is not designed to remove ammonia. The influent TKN is 30 mg/L and the final effluent TKN is 21 mg/L. Where did the 9 mg/L of TKN end up?
    a. Released to atmosphere
    b. Converted to ammonia
    c. Precipitated with alum
    d. Incorporated to biomass

7. The laboratory reports the following TKN and ammonia concentrations on a filtered and unfiltered sample of influent wastewater:

Unfiltered TKN as N	36.0 mg/L
Filtered TKN as N	26.0 mg/L
Ammonia-nitrogen (NH_3-N)	22.0 mg/L

Calculate the concentrations of particulate, soluble, and total organic nitrogen.

Chemistry Review

1. Which of the following terms refers to a basic building block of nature that can't be broken down further without losing its essential properties?
   a. Atom
   b. Element
   c. Compound
   d. Molecule

2. Elements are defined by
   a. Chemical properties
   b. Number of neutrons
   c. Chemical symbol
   d. Number of protons

3. The atomic number is the number of protons, but the atomic weight is
   a. Sum of protons, neutrons, and electrons
   b. Grams per molecule of diatom
   c. Sum of protons and neutrons
   d. Number of neutrons only

4. Molecules differ from compounds in this way.
   a. Compounds contain two or more atoms of the same element.
   b. Molecules are made from at least three different elements.
   c. Molecules must contain either oxygen or hydrogen.
   d. Compounds are made from at least two different elements.

5. Find the formula weight for calcium carbonate ($CaCO_3$). Calcium weighs 40 g/mole, carbon weighs 12 g/mole, and oxygen weighs 16 g/mole.
   a. 68 g/mole
   b. 84 g/mole
   c. 100 g/mole
   d. 204 g/mole

6. Find the formula weight for calcium nitrate [$Ca(NO_3)_2$]. The atomic weights are calcium = 40 g/mole, nitrogen = 14 g/mole, and oxygen = 16 g/mole.
   a. 102 g/mole
   b. 116 g/mole
   c. 164 g/mole
   d. 204 g/mole

7. What is the chemical formula for ferric chloride?
   a. $FeCl_2$
   b. $FeCl_3$
   c. FeOHCl
   d. $FeOCl_2$

8. Sodium ion has a +1 charge. Nitrite ion has a −1 charge. What is the chemical formula for sodium nitrite?
   a. $Na(NO_2)_2$
   b. Na_3NO_2
   c. $NaNO_2$
   d. Na_2NO_2

9. If ferrous ions (Fe^{+2}) are combined with phosphate ions (PO_4^{-3}), the chemical formula for the resulting compound would be
   a. Fe_2PO_4
   b. $Fe_3(PO_4)_2$
   c. $FePO_4$
   d. $Fe_6(PO_4)_2$

10. Chemical reactions must
    a. Be balanced
    b. Generate heat
    c. Produce acid
    d. Consume alkalinity

11. Which pH range will have the highest percentage of ammonia (NH_3) when the total ammonia concentration is 20 mg/L as N?
    a. <5
    b. 5 to 7
    c. 7 to 9
    d. >9

Biological Nitrification

1. Nitrification is a three-step process involving two groups of bacteria.
    ☐ True
    ☐ False

2. The NOB obtain their energy from nitrite and their carbon from alkalinity.
    ☐ True
    ☐ False

3. Carbonate and bicarbonate ions are two components of alkalinity in wastewater.
    ☐ True
    ☐ False

4. The AOB grow faster than the NOB when the water temperature is below 25 °C (77 °F).
    ☐ True
    ☐ False

5. In a well-functioning nitrification process, the effluent nitrite-nitrogen should be less than 1 mg/L as N.
    ☐ True
    ☐ False

6. The AOB obtain their energy from
    a. Ammonium ion (NH_4^+)
    b. DO (O_2)
    c. Alkalinity as $CaCO_3$
    d. Ammonia (NH_3)

7. pH affects the AOB by
    a. Binding to the cell membrane
    b. Reducing the availability of NH_3
    c. Interfering with oxygen uptake
    d. Increasing alkalinity

8. At pH 7 and 20 °C (68 °F), the total ammonia concentration was reported as 20 mg/L as N. What is the free or un-ionized ammonia concentration?
    a. Less than 0.2 mg/L
    b. 10 mg/L
    c. 13 mg/L
    d. 20 mg/L

9. Nitrifying bacteria convert ammonia to nitrate under
    a. Anaerobic conditions
    b. Anoxic conditions
    c. Aerobic conditions
    d. Reducing conditions

Process Variables for Nitrification

1. The operator of an activated sludge process knows from past experience that the water temperature in the aeration basin will drop from 22 °C (72 °F) to 15 °C (59 °F) between August and December. What process change should be made?
   a. Decrease the concentration with water temperature
   b. Increase the $SRT_{aerobic}$ gradually based on water temperature
   c. Increase the SRT and MLSS concentration to build biomass before November
   d. Wait until mid-October and then double the SRT

2. The minimum $SRT_{aerobic}$ needed for nitrification at 10 °C (50 °F) is 5 days. If a safety factor of 2.5 is used, what is the target $SRT_{aerobic}$?
   a. 2.5 days
   b. 5 days
   c. 12.5 days
   d. 25 days

3. An RBC process consists of four wheels operated in series. Where in the process is nitrification most likely to take place?
   a. First wheel
   b. Second wheel
   c. Third wheel
   d. Fourth wheel

4. A trickling filter receives secondary effluent that normally contains less than 15 mg/L of soluble BOD_5 and TSS. The trickling filter effluent normally contains less than 1 mg/L of NH_3-N. A process upset causes settled solids to escape the clarifier and enter the trickling filter. Which of the following is most likely to occur?
   a. Effluent ammonia concentrations will be unaffected.
   b. Effluent ammonia concentrations will increase.
   c. Effluent alkalinity concentrations will decrease.
   d. Effluent alkalinity concentrations will be unaffected.

5. A trickling filter receives dilute primary effluent containing 75 mg/L BOD_5 and 22 mg/L of NH_3-N. The effluent from the trickling filter contains 5 mg/L BOD_5 and <1 mg/L NH_3-N. Where in the trickling filter did nitrification begin?
   a. In the outer, aerobic layers of the biofilm
   b. Where DO concentrations reached 7 mg/L
   c. Where soluble BOD_5 decreased to <15 mg/L
   d. About two-thirds from the top of the media bed

6. An activated sludge process with an SRT of 6 days receives an abnormally high ammonia load. The blowers are unable to deliver enough air and the DO concentration in the basins drops to 1 mg/L. Which of the following may occur next?
   a. Nitrifying bacteria will die.
   b. Effluent ammonia will decrease.
   c. Average floc size will increase.
   d. Chlorine consumption will increase.

7. How much alkalinity will be needed to convert 20 mg/L ammonia-nitrogen (NH_3-N) to nitrate-nitrogen (NO_3-N)? There must be 100 mg/L of alkalinity as $CaCO_3$ remaining in the final effluent.
   a. 91 mg/L
   b. 143 mg/L
   c. 192 mg/L
   d. 243 mg/L

8. When supernatant from an anaerobic digester is returned to the activated sludge process in the middle of the day
   a. Effluent ammonia concentrations may increase.
   b. Alkalinity addition will be required to prevent pH drop.
   c. Nitrite-lock will occur and increase chlorine demand.
   d. Nitrifier growth rates will increase to meet the load.

9. This process control parameter is the least important in sustaining nitrification in activated sludge processes.
   a. Temperature
   b. SRT
   c. DO
   d. HDT

10. An activated sludge facility operates with a 25-day SRT in the winter. As soon as the water temperature drops below 16 °C, the SVI increases and foam accumulates on the activated sludge basins. The final effluent contains less than 1 mg/L NH_3-N. The chief facility operator begins to chlorinate the RAS to control filaments and improve settling. After 3 days, the effluent NH_3-N concentration increases to 22 mg/L. What must be true?
    a. Chlorine concentration in the RAS too high
    b. $SRT_{aerobic}$ is at least 50% of the total SRT
    c. SRT is not long enough to support nitrification
    d. Oxygen levels were depleted by chlorine addition

Biological Denitrification

1. Nitrogen gas produced during denitrification
   a. Is discharged with the final effluent.
   b. Ends up in the atmosphere.
   c. Combines with organic material.
   d. Improves sludge settling in clarifiers.

2. For denitrification to occur, which of the following conditions must exist?
   a. Excess alkalinity must be present.
   b. pH must be lower than 6.5.
   c. DO concentration is high.
   d. Organic carbon is available.

3. Dissolved oxygen can inhibit denitrification at concentrations as low as
   a. 0.3 mg/L
   b. 0.8 mg/L
   c. 1.3 mg/L
   d. 1.8 mg/L

4. Activated sludge processes that denitrify
   a. Consume more alkalinity than those that don't.
   b. Generate less sludge than those that don't.
   c. Often have difficulty maintaining neutral pH.
   d. Increase nitrogen loading to solids handling processes.

5. For every 1 mg/L NO_3-N converted to nitrogen gas, this much alkalinity is recovered.
   a. 3.57 mg/L as $CaCO_3$
   b. 7.14 mg/L as $CaCO_3$
   c. 7.48 mg/L as $CaCO_3$
   d. 8.34 mg/L as $CaCO_3$

6. Denitrification of 1 mg/L of NO_3-N recovers the equivalent of _____ mg/L of DO in the treatment process.
   a. 1.73 mg/L
   b. 2.86 mg/L
   c. 3.57 mg/L
   d. 7.14 mg/L

Process Variables for Denitrification

1. About this much BOD_5 is needed to convert 1 mg/L of nitrate-nitrogen (NO_3-N) to nitrogen gas (N_2).
   a. 1 mg/L
   b. 4 mg/L
   c. 7 mg/L
   d. 10 mg/L

2. Primary clarifiers can affect denitrification by
   a. Converting influent total nitrogen to nitrate-nitrogen
   b. Poisoning the anoxic zone with oxygen
   c. Changing the BOD_5 to nitrate-nitrogen ratio
   d. Increasing particulate BOD_5 to the secondary process

3. A WRRF needs to add organic carbon for denitrification. The operators are concerned about flammability and want something they don't have to test before each use. Which of the following carbon sources is best suited to their requirements?
   a. Methanol
   b. Waste beer
   c. Cheese whey
   d. Acetic acid

4. Which of the following must occur first?
   a. Convert ammonia to nitrite
   b. Convert nitrite to nitrate
   c. Convert nitrate to nitrogen gas
   d. Convert nitrogen gas to nitrite

5. Denitrification rates will be highest at this location.
   a. Bottom of a trickling filter
   b. Pre-anoxic zone
   c. Secondary clarifier blanket
   d. Post-anoxic zone

6. The F/M is 0.25 g BOD_5/g MLVSS at 10:00 a.m. If the F/M increases to 0.50 g BOD_5/g MLVSS at 2:00 p.m., the denitrification rate will
   a. Decrease by 50%
   b. Remain the same
   c. Approximately double
   d. Be limited by nitrate

7. Denitrification is most likely to occur in the secondary clarifier blanket when
   a. Influent flows and loads are low.
   b. Primary clarifier performs better than average.
   c. Methanol is added to tertiary filters.
   d. It is the middle of the day.

8. The IMLR
   a. Removes sludge from the clarifier
   b. Returns nitrate to the anoxic zone
   c. Recycles nitrified effluent in an RBC
   d. Wastes sludge to the digester

9. The influent to an MLE activated sludge process contains 20 mg/L nitrate-nitrogen (NO_3-N) and the effluent contains 12 mg/L NO_3-N. If the internal recycle is increased from 200 to 400% of influent flow, effluent nitrate-nitrogen remains the same. What must be true?

    a. Too much oxygen recycled to the anoxic zone

    b. Alkalinity has dropped below 50 mg/L as $CaCO_3$

    c. Inadequate supply of BOD_5 for more denitrification

    d. Upstream nitrification process has failed

Process Control for Denitrification

1. An activated sludge process has an anaerobic zone, anoxic zone, and aerated zone. Mixed liquor suspended solids are pumped from the end of the aerated zone to the beginning of the anoxic zone. The recycle flow is equal to 200% of the influent flow. The RAS is set to 75% of the influent flow. The biggest source of DO in the anoxic zone is

    a. Influent

    b. Primary effluent

    c. IMLR

    d. RAS

2. An activated sludge basin has a DO concentration of 2 mg/L. The ORP is

    a. Between -250 mV and -400 mV

    b. Between -250 mV and -175 mV

    c. Between -50 mV and $+50$ mV

    d. Between $+50$ mV and $+200$ mV

3. Which of the following statements about ORP is true?

    a. ORP increases at the end of denitrification.

    b. ORP is affected by water quality and temperature.

    c. The fill solution for an ORP probe is unimportant.

    d. In-basin ORP probes should match laboratory ORP results.

4. The primary effluent going to an activated sludge process has a BOD_5 concentration of 140 mg/L and an ammonia-nitrogen concentration of 55 mg/L. If the discharge permit limit for nitrate-nitrogen is 5 mg/L, what must be true?

    a. Supplemental carbon source may be needed.

    b. ORP will decrease through the aerated zone.

    c. IMLR should be set to 400% of influent flow.

    d. Sludge age should be less than 5 days.

5. This marks the end of denitrification when ORP is used to monitor an SBR.

    a. DO elbow

    b. Ammonia valley

    c. Nitrate knee

    d. Fermentation

Chemical Phosphorus Removal

1. Calcium oxide (lime) is the most commonly used chemical for phosphorus precipitation.

    ☐ True

    ☐ False

2. Orthophosphate and combined phosphates may be removed through chemical precipitation with ferric chloride or aluminum sulfate.

    ☐ True

    ☐ False

3. The chemical formula for alum, $Al_2(SO_4)_3 \cdot 14H_2O$, indicates that water molecules make up part of its crystal structure.
    - ☐ True
    - ☐ False

4. The chemical symbol for iron is
    - a. In
    - b. Ir
    - c. Fe
    - d. Pb

5. Which of the following chemicals is expected to produce the greatest quantity of sludge when used to precipitate phosphorus?
    - a. Ferric chloride
    - b. Calcium oxide
    - c. Aluminum sulfate
    - d. Ferrous sulfate

6. A WRRF has low influent alkalinity and must remove ammonia and phosphorus. Which of the following chemicals would be the best choice for precipitating phosphorus?
    - a. Ferric chloride
    - b. Ferrous sulfate
    - c. Sodium aluminate
    - d. Aluminum sulfate

7. The stoichiometric dose of aluminum ion to phosphorus is 0.87 g Al^{+3}/g P. If the influent total phosphate concentration is 10 mg/L as P and contains 50% orthophosphate, calculate the stoichiometric dose for removable P.
    - a. 4.35 mg/L
    - b. 8.7 mg/L
    - c. 12.6 mg/L
    - d. 17.4 mg/L

8. Currently, the WRRF is required to meet an effluent phosphorus limit of 1.5 mg/L as P. The new discharge permit lowers the limit to 0.05 mg/L. The amount of chemical needed to meet the new limit will
    - a. Increase by 33%
    - b. Approximately double
    - c. Increase by 75%
    - d. Increase exponentially

9. When aluminum sulfate reacts with water, these are formed.
    - a. Insoluble phosphates
    - b. Hydroxide compounds
    - c. Divalent ions
    - d. Insoluble sulfides

10. How many kilograms (pounds) of ferric chloride does 100 kg (pounds) of a 35% ferric chloride solution contain?
    - a. 17.5
    - b. 35
    - c. 65
    - d. 70

11. This is one reason a WRRF may decide to use ferric chloride instead of alum for phosphorus precipitation.
    - a. More expensive
    - b. Control filaments
    - c. Steel mill nearby
    - d. Odor control

12. Compared to ferric chloride, ferrous sulfate is
    a. Less efficient at precipitating phosphorus
    b. Better at reducing hydrogen sulfide
    c. More flexible with respect to addition points
    d. Available at higher concentrations

13. A new and improved iron-based chemical, Iron Blast, is being touted by the local sales representative. Given the following information, find the stoichiometric dose for this new chemical as milligram per liter Iron Blast per milligram per liter P.

    Chemical formula $-$ $Fe(NO_3)_3$
    1.8 g Fe^{+3}/g P, 2.71 g Fe^{+2}/g P
    Fe = 56 g/mole, N = 14 g/mole, O = 16 g/mole, and P = 31 g/mole
    a. 3.8 mg/L
    b. 7.8 mg/L
    c. 6.8 mg/L
    d. 14 mg/L

14. Vivianite is formed in anaerobic digesters when _____ is added to the digester.
    a. Ferric chloride
    b. Ferrous chloride
    c. Either ferric or ferrous chloride
    d. Neither ferric nor ferrous chloride

15. When either alum or ferric chloride is added to wastewater
    a. Hydrogen sulfide is precipitated.
    b. pH levels increase.
    c. Sludge production decreases.
    d. Alkalinity is consumed.

Sludge Production

1. Jar testing indicates that a dose of 76 mg/L of alum is needed to precipitate phosphorus. How much alkalinity will be consumed?
    a. 38 mg/L as $CaCO_3$
    b. 43 mg/L as $CaCO_3$
    c. 76 mg/L as $CaCO_3$
    d. 152 mg/L as $CaCO_3$

2. Chemical precipitation of phosphorus in aerobic processes works best when the pH is
    a. Acidic
    b. Neutral
    c. Basic
    d. Equilibrated

3. Ferric phosphate ($FePO_4$) is least soluble at pH
    a. 2
    b. 5
    c. 6
    d. 9

4. Chemical addition should only be done in locations where the wastewater is
    a. Septic
    b. Aerated
    c. Well mixed
    d. Quiescent

5. A WRRF is able to add alum in three different locations: primary clarifier outlet, activated sludge inlet channel, and the activated sludge outlet box. To maximize contact time for phosphorus precipitation, where should ferric chloride be added?

    a. Directly to activated sludge basin

    b. Primary clarifier outlet

    c. Activated sludge inlet channel

    d. Activated sludge outlet box

6. One reason alum may be added upstream of the primary clarifier is to

    a. Precipitate hydrogen sulfide

    b. Convert polyphosphates

    c. Remove ammonia

    d. Improve clarifier performance

7. Adding chemicals for phosphorus precipitation after the secondary clarifier is advantageous for this reason.

    a. Separate sludge handling equipment needed

    b. Increases fine particulates in the final effluent

    c. Prevents alkalinity loss from affecting nitrification

    d. Prevents conversion of ferrous iron to ferric iron

8. A WRRF must meet an effluent phosphorus limit of 0.10 mg/L total phosphorus. What must be true?

    a. Chemicals are added only upstream of the primary clarifier.

    b. Effluent suspended solids are removed by filtration.

    c. A combination of biological and chemical removal is required.

    d. MLSS total phosphorus concentration is less than 0.5%

9. The effluent TSS concentration is 10 mg/L. The MLSS is 4% phosphorus by weight. The influent contains 0.5 mg/L as P of soluble, nonbiodegradable condensed phosphates. Assuming all other phosphorus has been removed, what is the concentration of total phosphorus in the final effluent?

    a. 0.4 mg/L as P

    b. 0.9 mg/L as P

    c. 4.0 mg/L as P

    d. 4.5 mg/L as P

Biological Phosphorus Removal

1. Luxury uptake of phosphorus occurs

    a. In the anaerobic zone

    b. When the ORP is -250 mV

    c. In the presence of nitrate or oxygen

    d. When the SRT is less than 3 days

2. Volatile fatty acids include

    a. Glucose

    b. Acetic acid

    c. Sulfuric acid

    d. Carbohydrates

3. Which of the following biochemical reactions take place in the anaerobic zone?

    a. Consumption of PHB

    b. Storage of glycogen

    c. Release of phosphorus

    d. Secondary release

4. In an EBPR process, where will the liquid phase phosphorus concentration be the highest?
    a. Anaerobic zone
    b. Anoxic zone
    c. Aerated zone
    d. Secondary clarifier

5. Growth and reproduction of PAOs primarily takes place
    a. During VFA uptake
    b. When PHB is consumed
    c. During phosphorus release
    d. When glycogen is consumed

6. Phosphorus that is removed from an EBPR process
    a. Volatilizes to the atmosphere
    b. Combines with aluminum or iron
    c. Is sequestered in the WAS
    d. Occurs only under anoxic conditions

7. This organism is capable of both fermentation and luxury uptake of phosphorus.
    a. *Accumulibacter*
    b. *Tetrasphaera*
    c. *Competibacter*
    d. *Defluviicoccus*

8. About how long does it take for the PAOs to complete uptake of VFAs when they are in the anaerobic zone?
    a. Less than 10 minutes
    b. 15 to 45 minutes
    c. 90 to 120 minutes
    d. Longer than 3 hours

9. An extended aeration activated sludge process is operated with an SRT of 15 days in the winter. The new supervisor decides to increase the SRT to 35 days. How will this decision affect final effluent quality?
    a. Phosphorus concentrations may increase.
    b. Ammonia concentrations will decrease.
    c. Nitrite concentrations will increase.
    d. Orthophosphorus concentrations may decrease.

10. Phosphorus-accumulating organisms use VFAs in the anaerobic zone to do which of the following?
    a. Grow
    b. Gain energy
    c. Convert to internal storage products
    d. Reduce phosphorus

11. Which one of these is an indicator of good EBPR performance?
    a. High phosphorus content of MLSS
    b. Low BOD_5 concentration in the effluent
    c. Increased oxygen consumption in aerobic zone
    d. Negative ORP readings in the anoxic zone

12. Which ions are released in the anaerobic zone and taken up with phosphorus in the anoxic/aerated zones of an EBPR system?
    a. Magnesium and calcium
    b. Sodium and chloride
    c. Magnesium and potassium
    d. Bicarbonate and hydrogen

CHAPTER 10
Disinfection

Purpose and Function

1. Disinfection kills or inactivates all of the microorganisms in wastewater.
    - ☐ True
    - ☐ False

2. Chlorine works by destroying proteins and inactivating critical enzymes so microorganisms can't reproduce.
    - ☐ True
    - ☐ False

3. Why is disinfection of WRRF effluent required?
    a. To protect public health and the environment
    b. To decrease the potential for biological growth
    c. To reduce nutrient loading
    d. To completely eliminate all microorganisms

4. Facilities that use chlorine for disinfection
    a. Add sodium to neutralize chlorine and make it less toxic prior to discharge
    b. Have concentration limits for chlorine in their discharge permits
    c. Are required to maintain an effluent residual of at least 0.05 mg/L
    d. Discharge greater numbers of microorganisms than facilities using UV disinfection

5. One reason a WRRF may decide to use UV irradiation instead of chlorine for disinfection might be
    a. Chlorine is extremely expensive.
    b. Concerns over chlorine safety.
    c. Frequent power outages.
    d. Safety concerns with UV irradiation.

6. These compounds can form when chlorine reacts with dissolved organic compounds:
    a. Chloramines
    b. Organoozones
    c. Disinfection byproducts
    d. Sulfite salts

Indicator Organisms

1. If the final effluent from a WRRF contains 120 CFU/100 mL of *E. coli*, pathogens are present.
    - ☐ True
    - ☐ False

2. Reliable test methods exist for all of the pathogens potentially present in wastewater.
    - ☐ True
    - ☐ False

3. Indicator organisms
    a. Are ubiquitous (found everywhere)
    b. Require multiple testing methods
    c. Are associated with fecal contamination
    d. Indicate pathogens are present

4. Limits on which of the following microorganisms are typically included in discharge permits as an indicator of adequate disinfection?
    a. *Klebsiella*
    b. *E. coli*
    c. Cholera
    d. *Cryptosporidium*

5. Put these microorganisms in order from least specific (most inclusive group) to most specific.
    a. Fecal coliforms
    b. *E. coli*
    c. Total coliforms

6. The final effluent result for *E. coli* was reported as 350 MPN/100 mL. What must be true?
    a. The sample was filtered during analysis.
    b. The fecal coliform result must be less than 350 MPN/100 mL.
    c. Results include *Klebsiella* and other indicator organisms.
    d. A statistical table was used to estimate the number.

7. A vial from the MPN test is cloudy from bacterial growth, shows gas production, and glows blue under a black light. Which of the following statements is true?
    a. The test is positive for fecal coliforms.
    b. The test is positive for total coliforms.
    c. The test is positive for total coliforms and *E. coli*.
    d. The test is positive for total coliforms and fecal coliforms.

Chlorine Disinfection

1. Trade percent is the same thing as the percentage of available chlorine.
    ☐ True
    ☐ False

2. The chlorine concentration of all three forms of chlorine (gas, sodium hypochlorite solution, and calcium hypochlorite) is expressed as trade percent.
    ☐ True
    ☐ False

3. Dose is equal to the demand minus the residual.
    ☐ True
    ☐ False

4. Chlorine that has reacted with ammonia to form chloramines is still able to disinfect.
    ☐ True
    ☐ False

5. Chlorine and hypochlorites break down in water to form
    a. Hypochlorous acid and hypochlorite ion
    b. Hypochlorous acid and hydrochloric acid
    c. Hypochlorite ion and hydrogen ion
    d. Hydrochloric acid and chloride ion

6. The final effluent from a WRRF has extremely low alkalinity. Which of the following chemicals would NOT be a good choice for disinfection because it may cause the effluent pH to decrease?
   a. Chlorine gas
   b. Calcium hypochlorite
   c. Sodium hypochlorite
   d. Hypochlorous acid

7. Calcium chloride is used for disinfection. Which of the following statements is true?
   a. Alkalinity will be consumed and pH may decrease.
   b. Hypochlorous acid and hydrochloric acid will be produced.
   c. Alkalinity will be produced and pH may increase.
   d. Sodium hydroxide and hypochlorous acid will be produced.

8. At what pH will the percentage of hypochlorous acid be highest during chlorine disinfection?
   a. 6.5
   b. 7.0
   c. 7.5
   d. 8.0

9. The pH in the final effluent from a lagoon treatment facility rises during the day and falls at night. The chlorine dose is held constant throughout the day. The operator may need to
   a. Add alkalinity to minimize pH changes in the middle of the day
   b. Heat the final effluent to shift the chemistry toward hypochlorous acid
   c. Increase the dose to compensate for the shift toward hypochlorite ion
   d. Decrease the dose because of increased microorganism kill at high pH

10. Calculate the chlorine demand using the following data.

    —Raw water flow is 1.9 ML/d (0.50 mgd)
    —Chlorinator dose is 3.0 mg/L
    —Chlorine residual is 1.8 mg/L

    a. 0.8 mg/L
    b. 1.2 mg/L
    c. 3.0 mg/L
    d. 4.8 mg/L

11. Which of the following chemical species will NOT be measured in the TCR test?
    a. Hypochlorous acid
    b. Hypochlorite ion
    c. Chloride ion
    d. Monochloramine

12. The final effluent from a WRRF typically contains less than 0.25 mg/L of nitrite-nitrogen (NO_2-N). An upset in the activated sludge process has caused the NO_2-N concentration to increase to 3 mg/L. How will this affect a chlorine disinfection process?
    a. The nitrite-nitrogen will consume an additional 15 mg/L of chlorine.
    b. The nitrite-nitrogen will shift the equilibrium toward hypochlorite ion.
    c. The nitrite-nitrogen will not affect disinfection efficiency in any way.
    d. The nitrite-nitrogen will decrease chlorine demand by 2.75 mg/L.

13. When the chlorine to ammonia-nitrogen (NH_3-N) ratio is greater than 8 to 1, what must be true?
    a. Only monochloramine will form.
    b. Free chlorine residual will be present.
    c. TCR is maximized.
    d. Disinfection cannot occur.

14. Wastewater operators must be familiar with the break point curve because
    a. The ratio of chlorine to ammonia in the effluent can change during the day.
    b. Accidental operation at or near the break point can result in disinfection failure.
    c. Operating to the right of the break point consumes excess chemical and increases operating costs.
    d. All of the above.

15. The chlorine residual concentration, and ability to disinfect, are lowest
    a. To the left of the break point
    b. At the hump, before combined residual destruction
    c. At the break point
    d. To the right of the break point

16. Which of the following statements about break point chlorination is false?
    a. It is possible to read the same TCR concentration at more than one point on the break point curve.
    b. The combined chlorine residual is completely eliminated at the break point when enough chlorine is added.
    c. Wastewater operators sometimes operate to the right of the break point where free chlorine residual is measured.
    d. The TRC concentration includes the combined residual and the free residual.

17. An operator is able to adjust both the chlorine dose and the CT at their facility. If the CT is doubled and all other parameters remain the same, the operator could
    a. Decrease the dose by half to achieve the same CT
    b. Increase the dose by half to achieve the same CT
    c. Decrease the dose by 30% to achieve the same CT
    d. Increase the dose by 30% to achieve the same CT

18. When sulfur dioxide or sulfite salts are added to water containing chlorine
    a. Chlorine is converted to hypochlorous acid.
    b. Chlorine is oxidized to chlorohydrogen ions.
    c. Chlorine is reduced to chloride.
    d. Chlorine is converted to hypochlorite ion.

Design Parameters

1. Contact time will be higher at peak hour flow than at average daily flow.
    ☐ True
    ☐ False

2. The chlorine dose should be adjusted
    a. To reduce *E. coli* below 2 CFU/100 mL
    b. To meet discharge permit limits
    c. To match the design dose in the facility design manual
    d. To comply with Ten States Standards

3. The chlorine disinfection system is designed to provide 15 minutes of CT at peak hour flow. Which of the following statements must be true?
    a. Only one of the chlorine contact chambers needs to be in service.
    b. Chlorine doses must be increased to meet the minimum CT during average flow.
    c. There is enough CT to satisfy the design criteria in all states.
    d. Most of the time, the CT will be greater than 15 minutes.

4. U.S. EPA recommends that the *E. coli* concentration in recreational waters be kept, on average, below
    a. 2 CFU/100 mL
    b. 53 CFU/100 mL
    c. 126 CFU/100 mL
    d. 410 CFU/100 mL

5. Calculate the geometric mean for the following *E. coli* results: 2, 35, 1800, 45, 160.

    a. 62

    b. 76

    c. 98

    d. 408

Chlorination and Dechlorination Chemicals

1. Chlorine equipment and piping must be free of oil and grease.

    ☐ True

    ☐ False

2. As water temperatures increase, the amount of chlorine gas that can be dissolved in the water also increases.

    ☐ True

    ☐ False

3. Wet chlorine contains more than 150 mg/L of water.

    ☐ True

    ☐ False

4. Chlorine gas _____ than air.

    a. Contains more water

    b. Is heavier than

    c. Contains less water

    d. Is lighter than

5. This is an effective disinfectant that is less hazardous than chlorine or calcium hypochlorite.

    a. Sodium hypochlorite

    b. Potassium hydroxide

    c. Hydrochloric acid

    d. Sodium hydroxide

6. Sodium hypochlorite may _____ with time unlike gaseous chlorine.

    a. Degrade

    b. Crystalize

    c. Volatilize

    d. Strengthen

7. Sodium hypochlorite can have a short shelf life. Concentration will be affected by all but

    a. Temperature

    b. Time

    c. Concentration

    d. Pressure

8. The term *wet chlorine* refers to

    a. Sodium hypochlorite solutions

    b. Damp calcium hypochlorite powder

    c. Liquid chlorine containing free water

    d. Blends of calcium hypochlorite and water

9. This type of piping material should never be used with chlorine solutions (sodium hypochlorite)

    a. PVC

    b. Steel or iron

    c. PE

    d. FRP

10. A hypochlorite solution that is 15% (trade percent) strength contains this much chlorine:
    a. 1.5 g/L
    b. 15 g/L
    c. 150 g/L
    d. 1500 g/L

11. While chlorine smells like bleach, sulfur dioxide smells like
    a. Rotten eggs
    b. A freshly struck match
    c. Cabbage
    d. Decaying grass

12. Match the chemical to its color.

Chlorine gas	yellow
Chlorine liquid	colorless
Sulfur dioxide gas	white
Sodium bisulfite solution	greenish-yellow
Sodium metabisulfite	amber

Chlorine Containers

1. Fusible plugs are designed to melt at 70 to 74 °C (158 to 165 °F) to relieve gas pressure.
    ☐ True
    ☐ False

2. Fusible plugs only work when the cylinder valve is fully closed.
    ☐ True
    ☐ False

3. Rail cars may be used to supply either chlorine gas or chlorine liquid directly to the process.
    ☐ True
    ☐ False

4. A 68-kg (150-lb) chlorine cylinder typically contains this much chlorine:
    a. 45 kg (100 lb)
    b. 54 kg (120 lb)
    c. 68 kg (150 lb)
    d. 85 kg (188 lb)

5. Chlorine cylinders and containers are filled to a maximum of _____ of their volume.
    a. 70%
    b. 80%
    c. 90%
    d. 100%

6. Which of the following chlorine cylinders is likely to have the highest gas pressure?
    a. Cylinder stored at 10 °C (50 °F)
    b. Cylinder containing 75% liquid chlorine
    c. Cylinder located outdoors on a hot day
    d. Cylinder with a closed metal to metal seal

7. A 907-kg (1-ton) chlorine container has this many fusible plugs:
    a. 1
    b. 3
    c. 6
    d. 9

8. An operator connects a chlorine line to the bottom valve of a 907-kg (1-ton) container expecting to remove liquid chlorine. Instead, chlorine gas comes out through the lower valve. They switch the line to the upper valve, but still pull chlorine gas instead of liquid. What is the most likely reason for both valves to produce gas?

   a. Eductor tube has broken off within the container.

   b. Container is positioned so both valves are below the liquid level.

   c. Container is less than 50% full of liquid chlorine.

   d. Excessive heat has converted all of the liquid chlorine to gas.

Gas Chlorine Systems

1. The amount of chlorine in the liquid versus gas state in a cylinder, container, or rail car is dependent on temperature.

   ☐ True

   ☐ False

2. The maximum amount of chlorine that may be withdrawn from a 68-kg (150-lb) cylinder is 18 kg/d (40 lb/d).

   ☐ True

   ☐ False

3. Heat blankets may be applied to sulfur dioxide cylinders and containers, but not to chlorine vessels.

   ☐ True

   ☐ False

4. Liquid chlorine is used instead of chlorine gas when slower withdrawal rates are required.

   ☐ True

   ☐ False

5. Chlorine gas is withdrawn from a 68-kg (150-lb) cylinder. Condensation is observed on the bottom half of the cylinder. What must be true?

   a. The chlorine within the cylinder is beginning to freeze.

   b. The cylinder is approximately half full of liquid chlorine.

   c. If the withdrawal rate is decreased, ice may form.

   d. There is too much air movement around the cylinder.

6. Ice formation on cylinders and containers should be avoided because

   a. It can damage the cylinder or container wall.

   b. It may interfere with the function of the fusible plug.

   c. It prevents heat from moving into the cylinder or container.

   d. It increases the gas withdrawal rate and generates pressure.

7. Which of the following types of chlorine systems is the safest?

   a. Gas pressure system

   b. Cylinder-mounted chlorinator

   c. Wall-mounted chlorinator without vacuum regulator

   d. Liquid chlorine system

8. This piece of equipment is used in combination with a wall-mounted chlorinator to keep the entire system under vacuum.

   a. Vacuum regulator

   b. Venturi

   c. Eductor tube

   d. PTFE washer

9. Chlorine injectors function based on

   a. Pressure

   b. Coriolis effect

   c. Vacuum

   d. Plug flow

10. A chlorine system consists of a wall-mounted chlorinator and cylinder-mounted vacuum regulators. A leak develops between the chlorinator and the venturi. What is the most likely outcome?
    a. Gas room will fill with chlorine gas.
    b. Chlorinator diaphragm will close.
    c. Venturi operation will stop.
    d. Rotameter ball will remain elevated.

11. For a venturi to function correctly:
    a. The opening between the gas line and the venturi must be closed.
    b. The throat must be adjusted to the widest possible setting.
    c. Chlorine dose and residual must be greater than 10 mg/L.
    d. Pressure and flowrate must be high enough to produce a vacuum.

12. Manifolds are equipped with drip legs to
    a. Protect the chlorinator from liquid chlorine
    b. Increase heat transfer and withdrawal rates
    c. Reduce the need for sulfonator blankets
    d. Replace the pigtail-type connectors

13. The water flowrate through the venturi in a gas eductor has decreased below the manufacturer's recommended minimum flowrate. What is the most likely outcome?
    a. Gas flowrate will increase.
    b. Vacuum will be lost.
    c. Reliquificiation of chlorine.
    d. Outgassing to atmosphere.

14. Chlorine gas systems that operate under pressure must use this type of connector:
    a. Clear plastic pigtail
    b. Monel-lined copper pigtail
    c. Schedule 80 PVC
    d. Aluminum spiral wound PVC

Sodium and Calcium Hypochlorite

1. When withdrawing chlorine or sulfur dioxide liquid from a cylinder or container, this piece of equipment is needed to convert the liquid to a gas:
    a. Vacuum regulator
    b. Vaporizer
    c. Eductor tube
    d. Rupture disk

2. The tubing downstream of a chlorine evaporator is covered with a light frost. This indicates that
    a. The temperature within the evaporator is too high.
    b. Air circulation around the chlorine container is too low.
    c. Moisture is escaping from around the burst disk.
    d. The liquid withdrawal rate exceeds the evaporation rate.

3. To prevent metal corrosion from occurring within evaporators
    a. Zinc or magnesium rods create a sacrificial anode.
    b. Anti-corrosion chemicals are added to the interior chamber.
    c. Electrical current is applied across the two chambers.
    d. Very pure water is used within the outer chamber.

4. Gas storage rooms may be used to house tools and spare parts
    a. When tools must be kept nearby and other storage is unavailable.
    b. Only during the summer when the doors may be propped open.
    c. Never. Gas storage and use areas should be dedicated rooms.
    d. At any time. There are no restrictions on gas room storage.

5. When placing a new sulfur dioxide cylinder on a scale, the scale starting weight should be set to
    a. The tare weight of the cylinder
    b. The tare weight of the cylinder plus 68 kg (150 lb)
    c. The manufacturer's recommended default
    d. The weight of the cylinder minus the tare weight

6. Gas cylinders should be restrained
    a. 1/4 of the way up from the bottom
    b. 1/3 of the way up from the bottom
    c. 1/4 of the way down from the top
    d. 1/3 of the way down from the top

7. In the event of a liquid chlorine leak from a 907-kg (1-ton) container
    a. Safely rotate the container so gas escapes instead.
    b. Use the monorail hoist to remove the container from the room.
    c. Obtain a patch from a class B kit to repair the liquid leak.
    d. The liquid will freeze rapidly, stopping the leak.

8. Safety equipment should be located
    a. Next to the exhaust vent
    b. Outside the gas room
    c. In the administration building
    d. Between the empty and full containers

9. Match the equipment to its function.

1.	Rupture disk	a.	Connect cylinders and containers to manifolds
2.	Magnesium rods	b.	Reduces gas pressure
3.	Pigtail and yoke	c.	Relieves pressure when chlorine is trapped between closed valves
4.	Trunnions	d.	Uses water velocity to create a vacuum to pull chlorine
5.	Vacuum regulator	e.	Allow 907-kg (1-ton) containers to be easily rotated.
6.	Venturi	f.	Sacrificial anodes in evaporator water bath. Prevent corrosion.

10. The pump suction intake on a hypochlorite system should be placed
    a. As close to the tank bottom as possible
    b. Near the top of the tank
    c. Above the bottom of the tank
    d. About mid-depth in the tank

11. A centrifugal pump is used to transfer sodium hypochlorite solution to a diffuser in a chlorine contact chamber. Over time, the flow output of the pump has gradually decreased even though the pump speed has remained the same. What could be the problem?
    a. pH is too low
    b. Air binding
    c. Corrosion of pump housing
    d. Blockage

12. Sodium hypochlorite leaks through piping seams
    a. May be detected with carbonate solution
    b. Are the result of high alkalinity
    c. Are indicative of internal scaling
    d. Appear as large, white deposits

13. Water moves through the chlorine contact chamber as plug flow to
    a. Reduce the overall size of the chamber
    b. Minimize chlorine contact time
    c. Reduce construction costs
    d. Reduce opportunities for short-circuiting

14. A WRRF has two chlorine contact chambers. Each is designed to provide 15 minutes of CT at the peak hour design flow. The facility currently receives about 35% of its design flow. What must be true?
    a. One contact chamber does not provide enough CT at the current flow.
    b. Both contact chambers should be placed into service.
    c. Only one contact chamber is needed at the current influent flow.
    d. A higher chlorine residual concentration is needed.

Process Variables

1. The water temperature has decreased from 20 °C (68 °F) to 10 °C (50 °F). If 20 minutes of CT were needed before, how many minutes of CT are needed now?
    a. 10 minutes
    b. 20 minutes
    c. 30 minutes
    d. 40 minutes

2. This type of water is the most difficult to disinfect:
    a. Cold, low pH
    b. Warm, low pH
    c. Cold, high pH
    d. Warm, high pH

3. Experience has shown that a chlorine residual of 2 mg/L provides adequate disinfection with 30 minutes of CT. If the CT is reduced to 10 minutes because of a wet weather event, how much chlorine residual will be needed to achieve the same level of disinfection?
    a. 2 mg/L
    b. 4 mg/L
    c. 6 mg/L
    d. 8 mg/L

4. Chlorine gas is used to disinfect the final effluent. If the dose is 13.5 mg/L and the effluent flow is 9.46 ML/d (2.5 mgd), how much chlorine will be used each day?
    a. 1.28 kg/d (2.81 lb/d)
    b. 1.83 kg/d (4.05 lb/d)
    c. 115 kg/d (252 lb/d)
    d. 128 kg/d (281 lb/d)

5. A WRRF disinfects with 12% sodium hypochlorite solution. The effluent flow averages 37.85 ML/d (10 mgd). If the dose is set at 6 mg/L, how much sodium hypochlorite solution will be needed each day?
    a. 18.9 kg/d (41.7 lb/d)
    b. 227 kg/d (500 lb/d)
    c. 1670 kg/d (3740 lb/d)
    d. 1892 kg/d (4170 lb/d)

6. Given the following information, find the chemical feed pump setting in mL/min. Effluent flowrate = 87.06 ML/d (23 mgd), chlorine dose = 7.5 mg/L, hypochlorite strength = 12% available chlorine, hypochlorite density = 1.168 g/cm³ (1167.57 mg/mL).
    a. 1617 mL/min
    b. 3235 mL/min
    c. 6470 mL/min
    d. 7213 mL/min

7. The chlorine dose at the head of the contact chamber is 5.7 mg/L. At the end of the contact chamber, the chlorine residual is 2.2 mg/L. If the flowrate is 4.54 ML/d (1.2 mgd), how many kilograms (pounds) per day of sulfur dioxide will be needed?
    a. 9 kg/d (22 lb/d)
    b. 26 kg/d (57 lb/d)
    c. 36 kg/d (79 lb/d)
    d. 72 kg/d (180 lb/d)

Process Control

1. Flow pacing the chlorine dose to the effluent flow will always maintain a constant chlorine residual.
    ☐ True
    ☐ False

2. The ORP is higher after chlorination and lower after dechlorination.
    ☐ True
    ☐ False

3. Residual control adjusts the chlorine dose by measuring residual either directly or with ORP.
    ☐ True
    ☐ False

4. The simplest control strategy for chlorine disinfection is
    a. Manual
    b. Semiautomatic
    c. Flow-proportional
    d. Cascading loop

5. The electronic signal for an electronically controlled valve currently reads 4 mA. The valve is
    a. Fully closed
    b. 4% open
    c. 40% open
    d. Fully open

6. One advantage of compound-loop control is
    a. Slow reaction time
    b. Ability to manually adjust
    c. Steady chlorine residual
    d. Simplicity of operation

Operation—Daily Operation

1. A chlorine cylinder or container that is hot to the touch may be contaminated with internal moisture.
    ☐ True
    ☐ False

2. Cylinders may be rolled on their edges for transport so long as the distance traveled is less than 10 ft.
    ☐ True
    ☐ False

3. Cylinders and containers should be allowed to rest for 24 hours after delivery before being placed into service.
   ☐ True
   ☐ False

4. Lead washers may be reused if they are compressed sufficiently to form new grooves.
   ☐ True
   ☐ False

5. There is a federal law requiring SCBAs to be worn when changing chlorine cylinders and containers.
   ☐ True
   ☐ False

6. A gas cylinder valve should only be opened _____ turn, so it can be closed quickly in an emergency.
   a. One-quarter
   b. One-half
   c. One full turn
   d. Completely open

7. Chlorine leaks can be located quickly by
   a. Applying a dilute ammonia solution directly to a chlorine valve or fitting
   b. Using the vapor only from a squirt bottle filled with ammonia solution
   c. Spreading a thin layer of soapy water over the suspected leak and watching for bubbles
   d. Using a hand-held sulfur dioxide monitor calibrated for chlorine release

8. One method of confirming that there is no gas flowing through a chlorinator or sulfonator is to
   a. Check the downstream system pressure
   b. Confirm the scale weight equals tare weight
   c. Apply ammonia vapor at the valve stem
   d. Verify the ball is at the bottom of the rotameter

9. Clear plastic pigtails should be replaced when
   a. A crinkling sound is heard.
   b. They become discolored.
   c. A new cylinder or container is used.
   d. Manifolds are backflushed.

10. Deposits that form within the glass rotameter may
   a. Cause the float to become stuck
   b. Be removed with sodium hydroxide
   c. Interfere with pigtail operation
   d. Obscure the feed rate markings

11. This chemical may be used to remove calcium carbonate scale from a chemical-metering pump pumping sodium or calcium hypochlorite solution.
   a. Sodium hydroxide
   b. Sulfuric acid
   c. Hydrochloric acid
   d. Magnesium hydroxide

12. Chlorine contact chambers should be drained and cleaned at least
   a. Weekly
   b. Monthly
   c. Annually
   d. Every 2 years

Safety Considerations

1. Most humans are able to detect extremely low levels of chlorine gas by smell.
   - ☐ True
   - ☐ False

2. Sulfur dioxide is more toxic to humans than chlorine gas.
   - ☐ True
   - ☐ False

3. The OSHA has not set an exposure limit for sodium hypochlorite solutions.
   - ☐ True
   - ☐ False

4. The EPCRA requires facilities with at least _____ of chlorine gas on site to comply with reporting requirements and other provisions.
   - a. 45 kg (100 lb)
   - b. 68 kg (150 lb)
   - c. 272 kg (600 lb)
   - d. 907 kg (1 ton)

5. A chlorine release to the environment as small as _____ must be reported under EPCRA.
   - a. 1.0 kg (2.2 lb)
   - b. 4.5 kg (10 lb)
   - c. 45 kg (100 lb)
   - d. 68 kg (150 lb)

6. Under the Process Safety Management Rule, facilities with at least _____ lb of chlorine on-site have emergency planning requirements.
   - a. 453 kg (1000 lb)
   - b. 681 kg (1500 lb)
   - c. 1000 kg (2204 lb)
   - d. 1500 kg (3306 lb)

7. One goal of EPCRA is to
   - a. Reduce the amount of chlorine kept on-site.
   - b. Increase paperwork for utilities.
   - c. Keep first responders safe.
   - d. Ensure compliance with secondary containment rules.

8. Both the ACGIH and the OSHA set the safe level of exposure to chlorine gas over an 8-hour work day at
   - a. 0.1 ppmv
   - b. 0.5 ppmv
   - c. 1.0 ppmv
   - d. 10 ppmv

UV Disinfection

1. Visible light is the portion of the electromagnetic spectrum that can be seen by humans.
   - ☐ True
   - ☐ False

2. Some wavelengths of light can damage genetic material.
   - ☐ True
   - ☐ False

3. Damaged DNA can sometimes be repaired by a special enzyme within the cell.
    ☐ True
    ☐ False

4. Which of the following wavelengths of light would be most damaging to a bacterium?
    a. 190 nm
    b. 260 nm
    c. 480 nm
    d. 700 nm

5. Ultraviolet light disrupts DNA by
    a. Blasting apart the two halves of the double helix
    b. Substituting adenine for thymine in bonds
    c. Causing the formation of thymine dimers
    d. Directly interfering with base pair repair

6. Ultraviolet light is generated
    a. When liquid mercury fluoresces
    b. As a byproduct of static electricity
    c. Only at a wavelength of 254 nm
    d. By passing electricity through mercury vapor

Design Parameters

1. Ultraviolet disinfection may use either a high dose and short CT or a low dose and longer CT to achieve the same level of disinfection.
    ☐ True
    ☐ False

2. Quartz sleeves are used with UV lamps because they do not absorb UV light.
    ☐ True
    ☐ False

3. For chlorine disinfection, CT is the product of the chlorine residual concentration and DT. Which of the following terms for UV disinfection is most like CT?
    a. Fluence
    b. DT
    c. Fluence rate
    d. Millijoules

4. Ultraviolet disinfection efficiency may be reduced by all of the following EXCEPT
    a. Fouling of the quartz sleeve
    b. Decrease in effluent turbidity
    c. Aging of the UV lamp
    d. Increase in effluent TSS

5. Which of the following statements is true?
    a. UV light intensity decreases with distance from the lamp.
    b. UV light easily penetrates particles to inactivate microorganisms.
    c. Dead spaces within a UV reactor increase the DT.
    d. Fluence rates requirements are determined by the manufacturer.

6. As effluent turbidity increases
    a. Fecal coliform concentrations remain unchanged.
    b. Less UV light will be needed to disinfect.
    c. UV light penetration will also increase.
    d. *E. coli* concentrations are likely to increase.

Equipment

1. Banks contain more UV bulbs than modules.
    ☐ True
    ☐ False

2. Mechanical wipers may be manually, hydraulically, or pneumatically operated.
    ☐ True
    ☐ False

3. The intensity sensor in a UV system reports the average light intensity over the entire system.
    ☐ True
    ☐ False

4. Effluents that are clear and colorless have low UV absorbance and high UV transmittance.
    ☐ True
    ☐ False

5. The ballast in a UV module
    a. Provides power to each lamp
    b. Is intended to hold the lamps in place
    c. Adds weight to reduce vibration
    d. Always contains six lamps

6. Mechanical wipers in UV systems
    a. Prevent water droplets from accumulating
    b. Reduce biofilm and algae growth
    c. Automatically soak lamps in citric acid
    d. Typically contain glass fiber squeegees

7. Match the UV system component to its function.

    a. Lamp   1. Indirect measurement of UV dose
    b. Ballast   2. Cleans lamps within UV channel
    c. Dip tank   3. Generates UV light
    d. Quartz tube   4. Removes scale from quartz tubes
    e. Intensity monitor   5. Controls lamp temperature
    f. Transmittance monitor   6. Measures lamp light output
    g. Air scour   7. Provides power to lamps

8. This type of UV lamp emits light over many different wavelengths.
    a. Low-pressure
    b. Amalgam
    c. Low-pressure, high-intensity
    d. Medium-pressure

9. Ultraviolet lamps that emit light at only 254 nm are
   a. Monochromatic
   b. Medium-pressure
   c. Polychromatic
   d. High-pressure

10. Low-pressure UV bulbs typically last for 8000 hours of operation, but doing this can shorten lamp life.
    a. Operating lamps continuously
    b. Adjusting the wavelength above 300 nm
    c. Turning lamps on and off frequently
    d. Increasing the wiper cycle frequency

11. A mercury amalgam lamp
    a. Generates light at multiple wavelengths.
    b. Contains about twice as much mercury as a low-pressure lamp.
    c. Requires less time to warm up than a low-pressure lamp.
    d. Is more sensitive to temperature changes than other lamps.

12. A UV channel may contain baffles or a perforated plate
    a. To create turbulence
    b. To keep solids in suspension
    c. To hold up the ends of UV lamps
    d. To increase the DT

13. Manufacturers typically provide _____ UV sensors.
    a. One per bank
    b. One per module
    c. Two per bank
    d. Two per module

14. A WWRF receives industrial discharges that cause fluctuations in their effluent turbidity. Which of the following locations is preferred for a UV sensor that will be used to adjust lamp output?
    a. As close to the lamp as possible
    b. Halfway between two lamps
    c. Some distance away from the lamps
    d. At the entrance to the UV channel

15. The UV intensity reading has been gradually decreasing over the past month. What is the most likely cause?
    a. Rags or debris caught on sensor
    b. Inorganic or biological fouling of sleeves
    c. Increased flowrates and power usage
    d. New quartz sleeves over lamps

16. This type of meter would be most effective for determining the effectiveness of UV lamp cleaning.
    a. Transmittance meter
    b. Thermographic sensor
    c. Intensity meter
    d. Turbidity meter

17. The water level in a UV channel
    a. Should be kept as high as possible to keep lamps submerged.
    b. Is typically controlled with a fixed or adjustable weir.
    c. Increases and decreases as the facility influent flow changes.
    d. May be varied from day to day to reduce algae growth.

18. One consequence of UV lamps operating in air instead of remaining submerged might be
    a. Operator exposure to UV light and possible skin burns
    b. Excessive cooling of the lamps and loss of radiation
    c. Undisinfected layer of water passing over the lamps
    d. Flooding of ballasts and electrical short-circuiting

19. Effluent pumps send secondary effluent to the UV disinfection system. The effluent pump fails. The weir on the UV channel keeps the lamps submerged. If the pump remains off, what is likely to happen?
    a. Lamps will be exposed to air.
    b. UV channel may overflow.
    c. Ballasts may short-circuit.
    d. Lamps may overheat.

20. Of the following possible alarm conditions for a UV disinfection system, which is severe enough that it might require automatic shutdown of the UV system?
    a. Low UV transmittance
    b. Lamp/ballast failure
    c. Low liquid level
    d. Lamp nearing end of life

Maintenance

1. This commonly used chemical in WRRFs is known to reduce UV transmittance.
    a. Sodium hypochlorite
    b. Ferric chloride
    c. Sodium hydroxide
    d. Citric acid

2. Two common methods for conserving energy in UV disinfection are
    a. Flow pacing and dose pacing
    b. Supplemental chlorine addition and flow pacing
    c. Bypassing wet weather flows and adding chlorine
    d. Reducing cleaning cycles and watching television

3. How often should UV lamp run time values be monitored?
    a. Daily
    b. Weekly
    c. Monthly
    d. Semiannualy

4. Ultraviolet systems should only be started after
    a. Feed pumps have been started.
    b. Ballasts are fully de-energized.
    c. Channel covers are in place.
    d. All lamps are completely submerged.

5. Ultraviolet lamps should be replaced
    a. When they reach a predetermined number of service hours.
    b. When excessive fouling occurs that cannot be removed.
    c. When algae growth penetrates the quartz sleeves.
    d. When the lamp intensity falls below 90%.

6. For reclaimed water systems, NWRI requires intensity sensors be calibrated
    a. Daily
    b. Weekly
    c. Monthly
    d. Annually

7. Intensity sensors must be cleaned frequently
    a. To ensure accuracy
    b. To remove scale
    c. To prevent drift
    d. All of the above

8. Which of the following chemicals is often used for large-scale cleaning of UV lamps?
    a. Reagent-grade sulfuric acid
    b. Food-grade citric acid
    c. Sodium hypochlorite
    d. Ferric or ferrous chloride

Troubleshooting

Examine the photographs from a poorly performing UV system. In your own words, describe two problems that can be seen and the effect each will have on effluent quality.

_____

_____

_____

_____

(Reprinted with permission by Indigo Water Group)

Data Collection, Sampling, and Analysis

1. Operators may safely look directly at UV lamps while they are operating.
   - ☐ True
   - ☐ False

2. Keeping UV channels covered reduces algae growth and operator exposure to UV light.
   - ☐ True
   - ☐ False

3. Welder flash or conjunctivitis occurs immediately after prolonged UV exposure.
   - ☐ True
   - ☐ False

4. Mercury containment kits should be used to clean up any mercury released from a broken UV lamp.
   - ☐ True
   - ☐ False

5. Samples collected for either indicator organisms or chlorine residual must be
   a. Grab samples
   b. Composite samples
   c. Collected in sterile containers
   d. Contain sodium thiosulfate

6. Chlorine residual samples must be analyzed
   a. As soon as reasonably practicable.
   b. By a state certified laboratory.
   c. Within 15 minutes of collection.
   d. After cooling to less than 10 °C (50 °F).

7. A small amount of sodium thiosulfate is added to sample bottles intended for *E. coli* analysis
   a. As a preservative to increase hold time.
   b. To neutralize residual chlorine.
   c. To sterilize the sampling container.
   d. As a way to prevent *E. coli* from replicating.

List of Acronyms

CFU	= colony forming unit		CT	= contact time
°C	= degrees Celsius		cu ft	= cubic feet
°F	= degrees Fahrenheit		CWA	= Clean Water Act
μ	= specific growth rate of microorganisms, represented by Greek letter μ (pronounced mew)		DAFT	= dissolved air flotation thickener
			DBP	= disinfection byproduct
			DMR	= discharge monitoring report
μm	= micrometer = 0.001 mm		DNA	= deoxyribonucleic acid
μ_{max}	= maximum growth rate of microorganisms when saturated (not limited) by substrate		DO	= dissolved oxygen
			DOT	= Department of Transportation (U.S.)
μS	= microsiemens		DT	= detention time
A^2/O	= anaerobic, anoxic, oxic activated sludge process		EBPR	= enhanced biological phosphorus removal
			effluent	= treated wastewater
ABC	= Association of Boards of Certification		EPCRA	= Emergency Planning and Community Right-to-Know Act
ABS	= acrylonitrile-butadiene-styrene			
ac	= acre		EPS	= extracellular polymeric substances
ACGIH	= American Conference of Governmental Industrial Hygienists		F/M	= food-to-microorganism ratio
			FAC	= free available chlorine
ADF	= average daily flow		FIT	= flow indicator transmitter
aerobic	= in the presence of oxygen		FOG	= fats, oils, and grease
AIT	= analysis indicator transmitter		FRP	= fiber-glass-reinforced plastic (may also mean fiber-glass-reinforced polyester)
anaerobic	= without oxygen, nitrite, or nitrate			
anoxic	= without oxygen, but in the presence of nitrite or nitrate		ft	= feet
			ft/sec	= feet per second
AOB	= ammonia-oxidizing bacteria		g	= gram
ASME	= American Society of Mechanical Engineers		gal	= gallon
			gpd	= gallons per day
atm	= atmosphere		gpd/cap	= gallons per capita per day (per person)
ATP	= adenosine triphosphate			
AWT	= advanced wastewater treatment		gpd/sq ft	= gallons per day per square foot
BFP	= belt filter press		gpm	= gallons per minute
BNR	= biological nutrient removal		gpm/sq ft	= gallons per minute per square foot
BOD	= biochemical oxygen demand		GSA	= Gould sludge age
BOD_5	= biochemical oxygen demand (5-day)		H_2O	= water
BrINClHOF	= mnemonic device for remembering the elements that form two atom molecules with themselves in nature		H_2S	= hydrogen sulfide
			ha	= hectare
			HAA	= haloacetic acid
$C_6H_{12}O_6$	= sugar		HCO_3^{-2}	= bicarbonate ion
CAC	= combined available chlorine		HDPE	= high-density polyethylene
CBOD	= carbonaceous biochemical oxygen demand		HDT	= hydraulic detention time
			HHR	= high-resolution redox
$CBOD_5$	= carbonaceous biochemical oxygen demand (5-day)		HLR	= hydraulic loading rate
			HMA	= hot mix asphalt
CCTV	= closed-circuit television		HRT	= hydraulic retention time
CFR	= Code of Federal Regulations		HTH	= high-test hypochlorite
cfs	= cubic feet per second		IDLH	= immediately dangerous to life of health
CFU/100 mL	= colony forming units per 100 milliliters		IMLR	= internal mixed liquor recycle
CI	= Chlorine Institute		in.	= inch
CIU	= categorical industrial user		influent	= raw wastewater entering the WRRF
cm	= centimeter		JHB	= Johannesburg activated sludge process
CO_2	= carbon dioxide			
COC	= chain-of-custody		kg	= kilogram
COD	= chemical oxygen demand		kg/ha·d	= kilograms per hectare per day
CPVC	= chlorinated polyvinyl chloride		kg/m^3	= kilograms per cubic meter

kPa	= kilopascal, a unit of pressure	NPDES	= National Pollutant Discharge Elimination System
K_s	= half-saturation coefficient or half-saturation constant	NTF	= nitrifying trickling filter
L	= liter	NTU	= nephelometric turbidity unit
L/cap·d	= liters per capita (per person) per day	NWRI	= National Water Resources Institute
L/d	= liters per day	O&M	= operations and maintenance
lb	= pound	O_2	= oxygen
lb/ac/d	= pounds per acre per day	OLR	= organic loading rate
lb/cu ft	= pounds per cubic foot	ORP	= oxidation–reduction potential
lb/1000 cu ft/d	= pounds per thousand cubic feet per day	OSHA	= Occupational Safety and Health Administration
lb/1000 sq ft/d	= pounds per thousand square feet per day	OTR	= oxygen-transfer rate
lb/gal	= pounds per gallon	OUR	= oxygen uptake rate
LOTO	= lockout/tagout	P	= phosphorus
m	= meter	PACl	= polyaluminum chloride, also shown as PAX
m/s	= meters per second	PAO	= phosphate-accumulating organism
m^3	= cubic meter	PE	= polyethylene
m^3/d	= cubic meters per day	PEL	= permissible exposure level
$m^3/(m^2 \cdot d)$	= cubic meters per square meters per day	PFRP	= process to further reduce pathogens
MCRT	= mean cell residence time	PHB	= poly-β-hydroxybutyrate
Me_{dose}	= metal salt dose	P_{ini}	= initial soluble phosphorus concentration
mg	= milligram	PLC	= programmable logic controller
mg/L	= milligrams per liter	PMDAS	= mnemonic device for remembering the order of mathematical operations when solving equations—"Poly Makes Donuts After School" stands for parenthesis and powers, multiply, divide, add, and subtract
mgd	= million gallons per day		
mil. gal	= million gallons		
mJ/cm^2	= millijoules per square centimeters		
ML/d	= megaliters per day		
mL/min	= milliliters per minute		
MLE	= modified Ludzack-Ettinger activated sludge process	PO_4	= phosphate
		PO_4^{-3}	= phosphate ion
MLSS	= mixed liquor suspended solids	PO_4-P	= phosphate as phosphorus (as P)
MLVSS	= mixed liquor volatile suspended solids	PP	= polypropylene
mm	= millimeter = 0.001 m	ppd	= pounds per day
MPN/100 mL	= most probable number per 100 milliliters	ppm	= parts per million
MSDS	= material safety data sheet	ppmv	= parts per million per volume of air
MUG	= 4-methylumbelliferyl-beta-D-glucuronide	PRV	= pressure-reducing valve
mV	= millivolts	psi	= pounds per square inch
mW/cm^2	= milliwatts per square centimeters	psig	= pounds per square inch gauge
$mW \cdot sec/cm^2$	= milliwatt·seconds per square centimeters	PSRP	= process to significantly reduce pathogens
N_2	= nitrogen gas	PTFE	= polytetrafluorethylene
NDMA	= nitrosodimethylamine	PVC	= polyvinyl chloride
NH_3	= ammonia or free ammonia	PVdC	= polyvinyldiene chloride
NH_3-N	= ammonia-nitrogen (as N)	RAS	= return activated sludge
NH_4^+	= ammonium ion	RBC	= rotating biological contactor
NH_4^+-N	= ammonium-nitrogen ion (as N)	RR	= respiration rate
NH_4Cl	= ammonium chloride	RW	= Redwood media
NIOSH	= National Institute of Occupational Health and Safety	S	= concentration of limiting nutrient or substrate
nm	= nanometer = 0.001 μm	S	= elemental sulfur
NO_2^-	= nitrite ion	SARA	= Superfund Amendment and Reauthorization Act
NO_2	= nitrite		
NO_2-N	= nitrite-nitrogen (as N)	sBOD	= soluble BOD
NO_3^-	= nitrate ion	SBR	= sequencing batch reactor
NO_3	= nitrate	SCADA	= supervisory control and data acquisition
NO_3-N	= nitrate-nitrogen (as N)	SCBA	= self-contained breathing apparatus
NOB	= nitrite-oxidizing bacteria	SDS	= safety data sheet (formerly material safety data sheet or MSDS)

SIU	= significant industrial user		TOC	= total organic carbon
SK	= flushing intensity (Spülkraft)		TP	= total phosphorus
SLR	= solids loading rate		TRC	= total residual chlorine
SO_4^{-2}	= sulfate ion		TS	= total solids
SOR	= surface overflow rate		TSS	= total suspended solids
SOUR	= specific oxygen uptake rate		TVSS	= total volatile suspended solids
SPDES	= State Pollutant Discharge Elimination System		TWA	= time-weighted average
SRB	= sulfate-reducing bacteria		U.S. EPA	= U.S. Environmental Protection Agency
SRT	= solids retention time		UFC	= Uniform Fire Code
$SRT_{aerobic}$	= aerobic solids retention time		UV	= ultraviolet (radiation)
SSA	= specific surface area		UVT	= ultraviolet transmittance
SSC	= settled sludge concentration		vector	= a transmitter of disease
SSV	= settled sludge volume		VF	= vertical flow media
SSV_{30}	= settled sludge volume *after 30 minutes*		VFA	= volatile fatty acid
SSV_5	= settled sludge volume *after 5 minutes*		VFD	= variable-frequency drive
STEL	= short-term exposure limit		VOC	= volatile organic compound
STV	= statistical threshold value		VSS	= volatile suspended solids
SU	= standard units, used for pH		WAS	= waste activated sludge
SVI	= sludge volume index		WET testing	= whole effluent toxicity testing
TDS	= total dissolved solids		WRF	= Water Research Foundation
THM	= trihalomethane		WRRF	= water resource recovery facility (formerly wastewater treatment plant, or WWTP)
TIN	= total inorganic nitrogen			
TKN	= total Kjeldahl nitrogen (pronounced kell-doll)		XF	= crossflow media
TLV	= threshold limit value		Y	= yield, g VSS/g substrate

Answer Keys

CHAPTER 1

COMPONENTS OF WASTEWATER

1. False. Natural treatment systems have a limited capacity to assimilate pollutants. Urban areas concentrate a large number of people and businesses into a small area and discharge them into one or a few locations. The result is reduced dissolved oxygen, fish kills, and other adverse effects.
2. True
3. True
4. False, most come from animal sources like bacon.
5. B
6. C
7. D
8. A
9. D
10. A
11. A
12. C

THE CLEAN WATER ACT

1. True
2. True
3. True
4. False. The Federal penalties shown in Table 1.2, including jail time, only apply when a person knowingly and willfully falsifies, conceals, or covers up a permit violation or misrepresents a permit violation with fraudulent statements.
5. False. Permit applications and required reports must be signed by an authority state-licensed individual of responsible charge, a principal executive officer, or ranking elected official.
6. True
7. B
8. A
9. D

LIQUID TREATMENT PROCESSES—PRELIMINARY AND PRIMARY TREATMENT

1. True
2. False. Trash racks have wide openings to remove trash and large debris.
3. A
4. D
5. B
6. A
7. B
8. C
9. A

LIQUID TREATMENT PROCESSES—SECONDARY TREATMENT

1. True
2. True
3. True
4. False. The purpose of secondary treatment is to increase the size of particles remaining in the influent after screening, degritting, and primary sedimentation. The biological solids grown during secondary treatment must be separated from the treated wastewater before discharge using a secondary clarifier or other solids separation process.
5. False. Lagoons are lined to prevent contamination of underlying groundwater. They may be lined with natural clay (bentonite) or with a synthetic liner.

6. True
7. False. Trickling filters are so named because the wastewater is trickled down over the media surface. Free space between pieces of media is required for air to penetrate and provide oxygen to the microorganisms that make up the biofilm.
8. True
9. False. Disinfection reduces the numbers of bacteria and pathogens in the final effluent, but it does not sterilize the wastewater. Sterilization is the complete destruction or inactivation of all living things.
10. B
11. B
12. B
13. C
14. B
15. A
16. C
17. C
18. B
19. 1 = A, 2 = D, 3 = B, 4 = E, 5 = C
20. 1 = C, 2 = E, 3 = B, 4 = D, 5 = A

SOLIDS TREATMENT PROCESSES

1. True
2. True
3. False. Biosolids are regulated under Section 503 of Title 40 of the Code of Federal Regulations.
4. True
5. Secondary clarifier, thickening, digestion, dewatering
6. A
7. C
8. A
9. A
10. C
11. A
12. D
13. B
14. 1 = C, 2 = D, 3 = B, 4 = E, 5 = A
15. C

CHAPTER 2

SOURCES OF WASTEWATER

1. False. The Pretreatment Act does not distinguish between commercial and industrial users. It only defines domestic and industrial wastewater.
2. True
3. True
4. D
5. A
6. D

INFLUENT CHARACTERISTICS

1. True
2. False. An influx of stormwater would likely decrease the influent temperature.
3. True
4. False. While some conductivity meters give a TDS result, it is not a true measurement. TDS is calculated by applying a factor to the conductivity measurement. Although it is possible to build a correlation between conductivity and TDS for a particular wastewater, the only way to accurately measure TDS is by drying down a sample of filtered wastewater and weighing the residue left behind.
5. C
6. B

7. D
8. A
9. B
10. A
11. B
12. C
13. B

CHAPTER 3

INTRODUCTION TO PRELIMINARY TREATMENT
1. False. Headworks may contain all or none of these depending on the size and complexity of the WRRF.
2. False. Flow measurement can come before either screening or grit removal.
3. B
4. C
5. D

SCREENING PURPOSE AND DESIGN
1. True
2. True
3. True
4. C
5. A
6. C. The amount of screened material doubles for every 13-mm (0.5-in.) reduction of clear opening size. Here, the size was reduced twice so the amount of screenings will go up by a factor of 4.
7. A
8. D = micro screen, A = fine screen, C = bar screen,
   B = trash rack

SCREENING EQUIPMENT
1. True
2. False. Mechanical cleaning reduces labor costs.
3. False. Fine screens capture more of everything than coarse screens, including organic material.
4. True
5. A
6. 1 = E, 2 = G, 3 = A, 4 = B, 5 = F, 6 = C, 7 = H, 8 = D
7. D
8. C

SCREENING PROCESS CONTROL, OPERATION, AND MAINTENANCE
1. False. Combined sewers bring both sewage and stormwater to the WRRF. After a storm, expect accumulated debris to be flushed from the collection system and into the WRRF. Cleaning frequency should increase.
2. True
3. B
4. B
5. A = 3, B = 1, C = 2, D = 6, E = 5, F = 4, G = 7 (B, C, A, F, E, D, G)
6. A

GRIT REMOVAL PURPOSE AND DESIGN
1. False. The volume of grit removed is highly dependent on the activities in the service area, condition of the collection system, type of collection system, and other factors.
2. True
3. C
4. A
5. D
6. B

GRIT REMOVAL EQUIPMENT

1. False. Velocity is calculated by dividing the flowrate by the cross-sectional area. Alternatively, velocity can also be calculated by dividing the distance traveled by time.
2. True
3. C. To solve this problem, divide the current velocity by the desired velocity. The current velocity is 0.8 m/s (2.6 ft/sec). The maximum desired velocity is 0.3 m/s (1 ft/sec). $0.8 \div 0.3 = 2.7$ basins in service. The minimum desired velocity is 0.24 m/s (0.8 ft/sec). $0.8 \div 0.24 = 3.3$ basins in service.
4. D
5. A
6. C
7. B
8. D

GRIT REMOVAL PROCESS CONTROL, OPERATION, AND MAINTENANCE

1. True
2. False. The pump run time should be coordinated with both the time needed to fill the hopper and the pumping time needed to empty the hopper.
3. False. The mechanism should be started as soon as the flights are covered with water. Waiting until the basin is full will result in excessive amounts of grit accumulating around the flights. When finally started, the mechanism may not be able to move at all.
4. B
5. A
6. C
7. B

ADDITIONAL PRETREATMENT CONSIDERATIONS

1. True
2. False. Septage tends to be higher strength than domestic wastewater.
3. True
4. False. Grease trap waste is considered industrial and commercial waste, not domestic. It is regulated under 40 CFR Part 257.
5. C
6. B
7. D

CHAPTER 4

INTRODUCTION TO PRIMARY TREATMENT OF WASTEWATER

1. False. The clarifier can only remove the settleable portion of TSS and BOD_5.
2. D
3. B

THEORY OF OPERATION

1. True
2. False. The particle with more surface area will settle slower, not faster.
3. False. Compaction is complete after about 2 hours.
4. C
5. D
6. B
7. C
8. A

DESIGN PARAMETERS

1. False. The percent removal depends on the amount of particulate and soluble BOD_5 entering the clarifier.
2. True
3. True
4. B
5. D
6. B

EQUIPMENT—CIRCULAR CLARIFIERS

1. True
2. True
3. False. Most clarifiers have 0.5 to 0.7 m (1.5 to 2.0 ft) of freeboard.
4. C
5. B
6. A
7. True
8. False. The feedwell reduces inlet velocity.
9. B
10. A

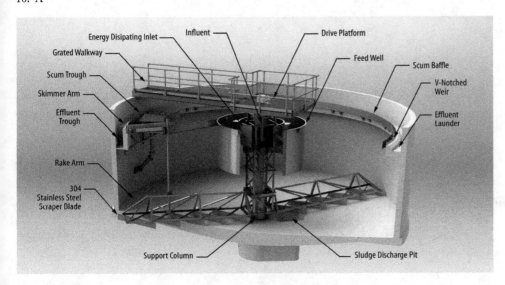

11. False. Weirs for circular clarifiers are typically v-notch weirs.
12. True
13. False. The drive is typically mounted on the center platform.
14. D
15. C
16. A
17. B
18. True
19. True
20. True
21. B
22. D
23. C
24. B
25. C
26. A

EQUIPMENT—RECTANGULAR CLARIFIERS

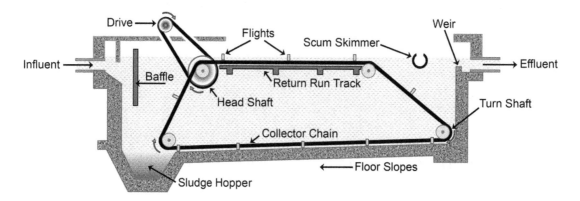

1. False. Flow is longitudinal from one end to the other.
2. True
3. True
4. B
5. D
6. D
7. A
8. B
9. C
10. A
11. D
12. C
13. C

PROCESS VARIABLES

1. True
2. False. *Surface overflow rate* is another name for velocity specific to clarifiers.
3. 4805.6 m³ (169 560 cu ft)
4. 1.8 hours
5. 28.8 m³/m²·d (707.7 gpd/sq ft)
6. D
7. A
8. A
9. C
10. D
11. B

PROCESS CONTROL—HYDRAULIC CONSIDERATIONS AND SOLIDS HANDLING

1. True
2. True
3. 1400 kg/d (3083 lb/d)
4. 63.7 m³/d (16 828 gpd)
5. 46.4 minutes
6. B
7. C
8. A
9. D
10. B

PROCESS CONTROL—SKIMMING, UPSTREAM, AND DOWNSTREAM UNITS

1. True
2. False. Recycle flows should be returned when they have the least effect. Returning during peak hour flow will increase the SOR, possibly to unacceptably high levels.
3. C
4. D
5. B

OPERATION

1. False. Some sludge collection mechanisms have water-lubricated bearings. These mechanisms should only be started when they are submerged.
2. True
3. False. Sludge left in lines will continue to break down. Gasses produced may build pressure and can damage equipment and lines.
4. C
5. A
6. D
7. B

MAINTENANCE

1. True
2. B
3. A
4. C

TROUBLESHOOTING

1. A
2. C
3. B
4. D

CHAPTER 5

INTRODUCTION TO BIOLOGICAL TREATMENT

1. True
2. False. A microscope is needed to see individual bacteria.
3. False. Some material is biodegradable, but takes longer than 5 days to break down and will not be measured by the 5-day test.
4. False. Liquid is absorbed, not adsorbed, into a sponge.
5. C
6. B
7. C
8. D
9. B
10. A

MICROBIOLOGY

1. True
2. True
3. True
4. False. Metazoa are larger than bacteria and protozoa.
5. B
6. D
7. B
8. C
9. A

BACTERIA IN SECONDARY TREATMENT PROCESSES

1. True
2. False. Autotrophic bacteria cannot use CBOD. They obtain carbon from dissolved carbonate.
3. False. Autotrophic bacteria use energy-poor fuel sources like ammonia.
4. True
5. C
6. B
7. 1 = C, 2 = B, 3 = A
8. C
9. B
10. C
11. A
12. B
13. C
14. C
15. B

MICROBIAL GROWTH RATES

1. True
2. True
3. False. The growth rate will be limited by the number of bacteria in the process.
4. False. Each substrate has a minimum concentration needed to support the fastest possible growth of the bacteria. Multiple substrates may be limiting. For example, nitrifying heterotrophic bacteria could lack both oxygen and ammonia.
5. True
6. C
7. B
8. D
9. B
10. D
11. A

CHAPTER 6

PURPOSE AND FUNCTION

1. False. Not all pond systems have screening and grit removal.
2. True
3. False. Ponds do not remove nutrients consistently.
4. C
5. D
6. A
7. D

THEORY OF OPERATION—FACULTATIVE PONDS

1. C
2. B
3. B
4. C
5. A
6. B
7. 1 = C, 2 = F, 3 = A, 4 = E, 5 = B, 6 = D
8. B

POND BIOLOGY

1. A
2. C
3. B
4. A
5. D
6. D
7. B
8. C
9. A
10. C
11. B
12. A
13. B
14. C
15. C

DESIGN PARAMETERS

1. B
2. C
3. C

EXPECTED PERFORMANCE

1. D
2. B
3. C
4. A

EQUIPMENT

1. False. Actual detention times are shorter because of incomplete mixing.
2. False. Dike slopes are not as steep on the outside of the pond. Steep slopes are hard to mow.
3. True
4. B
5. C
6. 1 = C, 2 = F, 3 = E, 4 = A, 5 = B, 6 = G, 7 = D
7. A
8. D
9. A
10. C
11. B
12. D
13. D
14. B
15. C
16. A
17. D

PROCESS VARIABLES

1. True
2. True
3. False. Oxygen saturation concentrations decrease with decreasing barometric pressure (elevation).
4. C
5. C
6. B
7. A
8. B

9. D
10. A
11. C

OPERATION

1. False. Aquashade and other dyes that block sunlight are not approved by U.S. EPA for use in wastewater treatment ponds.
2. True
3. B
4. B
5. D
6. A

MAINTENANCE

1. C
2. A
3. A
4. D
5. False. Regulations change over time. A chemical applied last season or last week may no longer be on the approved list.
6. True

CHAPTER 7

PURPOSE AND FUNCTION OF FIXED-FILM PROCESSES

1. True
2. True
3. False. Trickling filters are not submerged. The wastewater trickles over the media.
4. D
5. C
6. A

THEORY OF FIXED-FILM OPERATION

1. True
2. False. Substances move from areas of high concentration to areas of low concentration.
3. False. EPS increases with biofilm age.
4. A
5. B
6. C
7. A
8. B
9. D
10. B
11. D
12. C
13. C

TRICKLING FILTERS

1. True
2. True
3. False. As loading increases, as with a roughing filter, the percent removal decreases and the BOD concentration in the effluent increases.
4. C
5. C
6. B

7. A

8. B

9. C

TRICKLING FILTER EQUIPMENT

1. False. Many trickling filters are hydraulically driven and do not have motors.

2. True

3. A

4. C

5. D

6. B

BIOFILM SUPPORT MEDIA

1. A = 3, B = 5, C = 1, D = 2, E = 4

2. C

3. D

4. B

5. A

CONTAINMENT TANKS AND STRUCTURES

1. True

2. False. The bearings at the top of the mast are better protected from corrosion.

3. B

4. D

5. A

6. A

7. D

8. C

9. A

PROCESS VARIABLES FOR TRICKLING FILTERS

1. False. It is the mass of BOD_5 per VOLUME of trickling filter media.

2. True

3. D

4. A

5. A

6. D

7. C

8. A

9. D

10. B. The volume of the trickling filter is 9679.8 m³ (343 359 cu ft) and the total load of BOD going to the trickling filter is 4086 kg (9007.2 lb). The recycle flow should not be included in this calculation.

11. C

12. B

VENTILATION

1. False. Water will fall through the tower. Heat will be transferred from the hot air to the cooler water. This makes the air inside the trickling filter cooler than the outside air.

2. True

3. True

4. B

5. B

6. D

7. B

OPERATION OF TRICKLING FILTERS

1. True
2. False. Taking grit basins and primary clarifiers out of service can increase the amount of grit, grease, scum, and organic load going to the trickling filter. Excess grit can cause media plugging and grease and scum accumulation, especially in rock media trickling filters. Increasing the organic loading rate will increase the amount of biofilm produced and may require an adjustment of both the recycle ratio and distributor speed.
3. True
4. B
5. A
6. C
7. C
8. A

DATA COLLECTION, SAMPLING, AND ANALYSIS FOR TRICKLING FILTERS

1. B
2. C. Should also check to see if distributor orifices are blocked.
3. B. A hooked wire may also be used.
4. A
5. D
6. C
7. A

ROTATING BIOLOGICAL CONTACTORS

1. B
2. C
3. C
4. A
5. D
6. B
7. A

EQUIPMENT FOR ROTATING BIOLOGICAL CONTACTORS

1. True
2. False. Metals expand when heated and contract when cooled.
3. True
4. B
5. C
6. A
7. B
8. C
9. B
10. C
11. A

PROCESS VARIABLES FOR RBCs

1. False. Loading is calculated as mass applied per surface area.
2. False. Ammonia removal requires more time.
3. B
4. C
5. B
6. C
7. B

PROCESS CONTROL FOR ROTATING BIOLOGICAL CONTACTORS
1. B
2. C
3. B
4. D
5. A
6. C

OPERATION OF ROTATING BIOLOGICAL CONTACTORS
1. B
2. C
3. A
4. D
5. A
6. C
7. B

CHAPTER 8

PURPOSE AND FUNCTION
1. True
2. True
3. B
4. D
5. C
6. A

THEORY OF OPERATION
1. True
2. False. MLSS particles settle at the same speed as a blanket.
3. C
4. D
5. A
6. B
7. B
8. B
9. A
10. C
11. B

ACTIVATED SLUDGE MICROBIOLOGY
1. False. Too many filaments will cause settling problems in the clarifier.
2. A = 4, B = 1, C = 7, D = 2, E = 6, F = 3, G = 5
3. C
4. D
5. A = 3, B = 5, C = 1, D = 2, E = 4
6. A

DESIGN PARAMETERS
1. B
2. A
3. B
4. C
5. A

EQUIPMENT

1. True
2. True
3. False. The oxygen must be transferred from the air to the water to be dissolved.
4. D
5. B
6. C
7. D
8. B
9. C
10. A
11. B
12. D
13. D
14. A
15. B
16. A
17. B
18. D
19. B
20. C

PROCESS VARIABLES FOR THE ACTIVATED SLUDGE BASIN

1. True
2. True
3. D
4. A
5. C
6. D
7. B
8. C. [(2.3 mgd ÷ 1.45 mgd) + 1] × 2600 mg/L = 6724 mg/L, round to 6700 mg/L
9. B. (1000 mL ÷ 280 mL) × 1850 mg/L = 6607 mg/L
10. B. The clarifier can't perform better than the settleometer.
11. C
12. D
13. C. (2000 mL ÷ 280 mL) × 1800 mg/L = 12 857 mg/L
14. A. To solve, assume the influent flow is 1 m³/d (or 1 mgd) and set the RAS flow to 0.7 m³/d (or 0.70 mgd). Then, RAS concentration = [(1.0 ÷ 0.7) + 1] × 2600 mg/L = 6314 mg/L
15. A. 10 000 mg/L = [(1.0 ÷ RAS flow) + 1] × 2500 mg/L. Rearrange the equation to solve for RAS flow.

SLUDGE AGE

1. False. Sludge age is how long the average sludge particle remains in the system before being wasted.
2. True
3. A
4. A
5. B
6. B
7. A
8. B
9. B
10. D
11. B
12. D

13. B
14. C
15. D
16. B. First, find the mass of MLSS in the activated sludge basin. This is 28 387.5 kg (62 550 lb). Then, find the mass of WAS. This is 3488.3 kg (7686 lb). For U.S. customary units, it is necessary to convert the wasting rate in gpm to mgd first. 80 gpm = 0.1152 mgd. Then, divide the mass of MLSS by the mass of WAS to find SRT.

POTENTIAL EFFECTS OF FOAMING EVENTS ON SLUDGE AGE

1. False. Wasting should be done at least daily.
2. A
3. B
4. D
5. B
6. C
7. A
8. C
9. B

FOOD-TO-MICROORGANISM RATIO

1. True
2. A. First, find the total mass of MLSS. Then, multiply by 0.8 (80%) to find the mass of MLVSS.
3. B. If the clarifier removes 35%, then 65% of the load goes through to the activated sludge process.
4. B
5. D

RELATIONSHIPS BETWEEN PROCESS VARIABLES

1. False. Setting the sludge age also sets the F/M.
2. False. Yield is fixed. It isn't possible to grow more microorganisms without more load.
3. C
4. B
5. A
6. C
7. D
8. D
9. A = 2, B = 3, C = 5, D = 1, E = 4

PROCESS CONTROL FOR ACTIVATED SLUDGE BASINS

1. False. BOD uptake can be complete even if the MLSS is not flocculating.
2. True
3. Detention time = extended air is longer, SRT = extended air is longer, F/M = conventional is higher, volumetric loading rate = conventional is higher
4. A
5. C
6. C
7. A
8. D
9. B
10. B
11. D
12. C
13. C
14. B

SETTING A TARGET DISSOLVED OXYGEN CONCENTRATION
1. B
2. C
3. B
4. B
5. A
6. D

PROCESS VARIABLES FOR SECONDARY CLARIFIERS
1. D
2. B. The surface area of the clarifier is 308 m² (3317 sq ft)
3. C
4. B. The surface area of the clarifier is 1052 m² (11 304 sq ft)
5. C
6. A. The distance from the outer wall to the weir must be subtracted from both sides.
7. B. The RAS concentration is extra information not needed to solve the problem.
8. D
9. A
10. D
11. B

PROCESS CONTROL FOR SECONDARY CLARIFIERS
1. D
2. B
3. C
4. A
5. B
6. C

OPERATION—DAILY OPERATION
1. False. Seed sludge should come from a well-functioning activated sludge basin.
2. True
3. B
4. D
5. A
6. B

DATA COLLECTION, SAMPLING, AND ANALYSIS
1. A
2. B
3. B
4. C
5. B
6. D
7. B
8. C
9. B
10. A
11. C
12. D
13. D
14. B
15. C

16. C
17. C
18. B

MAINTENANCE

1. True
2. False. Take one corrective action at a time.
3. B
4. A
5. a, c, d, and e are all process control. b and f are troubleshooting.

AERATION SYSTEM TROUBLESHOOTING

1. B
2. A
3. D
4. B
5. B

SLUDGE QUALITY PROBLEMS

1. 1 = E, 2 = B, 3 = G, 4 = H, 5 = D, 6 = C, 7 = F, 8 = A
2. B
3. A
4. D
5. B
6. C
7. B
8. D
9. D
10. C
11. A

SECONDARY CLARIFIERS

1. True
2. B
3. C
4. D
5. B
6. A
7. B
8. C
9. A
10. B
11. C

CHAPTER 9

NITROGEN AND PHOSPHORUS IN WASTEWATER

1. B
2. A
3. C
4. A
5. C. 1 mg/L of P is removed for every 100 mg/L BOD_5. 300 mg/L BOD_5 = 3 mg/L P

6. D
7. The total organic nitrogen is the difference between the TKN and the ammonia-nitrogen:

$$\text{Organic nitrogen} = 36.0 - 22.0 = 14.0 \text{ mg/L N}$$

All of the ammonia is soluble. Therefore, the soluble organic nitrogen is the difference between filtered TKN and the ammonia concentration.

$$\text{Soluble organic nitrogen} = 26.0 - 22.0 = 4.0 \text{ mg/L N}$$

The particulate organic nitrogen is the difference between the total organic nitrogen and the soluble organic nitrogen.

$$\text{Particulate organic nitrogen} = 14.0 - 4.0 = 10.0 \text{ mg/L N}$$

CHEMISTRY REVIEW
1. A
2. D
3. C
4. D
5. C
6. C
7. B
8. C
9. B
10. A
11. D

BIOLOGICAL NITRIFICATION
1. False. Nitrification is a two-step process.
2. True
3. True
4. False. The NOB grow faster than the AOB when water temperatures are below 25 °C.
5. True
6. D
7. B
8. A
9. C

PROCESS VARIABLES FOR NITRIFICATION
1. B
2. C
3. D
4. B
5. C
6. D
7. D
8. A
9. D
10. A

BIOLOGICAL DENITRIFICATION
1. B
2. D
3. A
4. B
5. A
6. B

PROCESS VARIABLES FOR DENITRIFICATION
1. B
2. C
3. D
4. A
5. B
6. C
7. D
8. B
9. C

PROCESS CONTROL FOR DENITRIFICATION
1. C
2. D
3. B
4. A
5. C

CHEMICAL PHOSPHORUS REMOVAL
1. False. Calcium oxide (lime) usage is less common than both aluminum and iron salts.
2. False. Only orthophosphate can be chemically precipitated.
3. True
4. C
5. B
6. C
7. A
8. D
9. B
10. B
11. D
12. A
13. B
14. C
15. D

SLUDGE PRODUCTION
1. A
2. A
3. B
4. C
5. B
6. D
7. C
8. B
9. B

BIOLOGICAL PHOSPHORUS REMOVAL
1. C
2. B
3. C
4. A
5. B
6. C
7. B
8. B
9. A

10. C
11. A
12. C

CHAPTER 10

PURPOSE AND FUNCTION

1. False. Sterilization kills or inactivates all microorganisms. Disinfection reduces their numbers to safe levels.
2. True
3. A
4. B
5. B
6. C

INDICATOR ORGANISMS

1. False. Pathogens may or may not be present.
2. False. Many pathogens do not have reliable test methods to determine their presence.
3. C
4. B
5. C. Total coliforms → A. fecal coliforms → B. *E. coli*
6. D
7. C

CHLORINE DISINFECTION

1. True
2. False. Calcium hypochlorite is expressed as a weight ratio of chlorine.
3. False. Dose is equal to demand PLUS residual.
4. True
5. A
6. A
7. C
8. A
9. C
10. B
11. C
12. A
13. B
14. D
15. C
16. B
17. A
18. C

DESIGN PARAMETERS

1. False. More flow is moving through the facility at peak hour flow, which reduces the amount of time it spends in the process.
2. B
3. D
4. C
5. A

CHLORINATION AND DECHLORINATION CHEMICALS

1. True
2. False. Solubility decreases with increasing temperature.
3. False. How much water chlorine can contain before it is considered "wet" depends on temperature.
4. B
5. A

6. A

7. D

8. C

9. B

10. C

11. B

12. Chlorine gas = greenish-yellow, Chlorine liquid = amber, Sulfur dioxide gas = colorless, Sodium bisulfite solution = yellow, Sodium metabisulfite = white

CHLORINE CONTAINERS

1. True

2. False. The fusible plug is located below the metal to metal seal and can open regardless of the valve position.

3. False. Although rail cars are equipped with gas lines, they are used for pressurizing the car rather than removing chlorine gas.

4. C

5. B

6. C

7. C

8. A

GAS CHLORINE SYSTEMS

1. True

2. False. Although this is the recommended reliable withdrawal rate, higher rates may be obtained by minimizing backpressure and increasing air circulation.

3. True

4. False. Liquid chlorine is used when gas cannot be withdrawn from the container fast enough to meet demand.

5. B

6. C

7. B

8. A

9. C

10. B

11. D

12. A

13. B

14. B

SODIUM AND CALCIUM HYPOCHLORITE

1. B

2. D

3. A

4. C

5. D

6. D

7. A

8. B

9. 1 = C, 2 = F, 3 = A, 4 = E, 5 = B, 6 = D

10. C

11. B

12. D

13. D

14. C

PROCESS VARIABLES

1. D

2. C

3. C

4. D

5. D

6. B

7. A

PROCESS CONTROL

1. False. Changes in effluent quality can cause changes in the chlorine residual concentration even when the chlorine feed is flow paced.

2. True

3. True

4. A

5. A

6. C

OPERATION—DAILY OPERATION

1. True

2. False. A handtruck (dolly) should be used no matter how short the distance.

3. True

4. False. Lead washers may never be reused.

5. False. There is no requirement to wear an SCBA during changeout.

6. A

7. B

8. D

9. B

10. A

11. C

12. C

SAFETY CONSIDERATIONS

1. False. Most humans can't detect low levels of chlorine, which is why gas detectors are needed in gas rooms.

2. False. Sulfur dioxide is less toxic than chlorine.

3. True

4. A

5. B

6. C

7. C

8. B

UV DISINFECTION

1. True

2. True

3. True

4. B

5. C

6. D

DESIGN PARAMETERS

1. True

2. False. Quartz sleeves absorb some UV light.

3. C

4. B

5. A

6. D

EQUIPMENT

1. True. Banks are groups of modules.

2. True

3. False. The intensity sensor can only report the light intensity at the sensor location.

4. False. Clear, colorless effluents may contain UV absorbing substances.
5. A
6. B
7. A = 3, B = 7, C = 4, D = 5, E = 6, F = 1, G = 2
8. D
9. A
10. C
11. B
12. A
13. A
14. C
15. B
16. C
17. B
18. A
19. D
20. C

MAINTENANCE

1. B
2. A
3. C
4. D
5. A
6. C
7. D
8. B

TROUBLESHOOTING

The control panel shows multiple lamps are burned out. This creates a hydraulic hole. Water passing through these areas is not being disinfected. The image of the lamps shows an excessive accumulation of solids and large solids may be seen between the lamps. This reduces transmissivity and reduces disinfection.

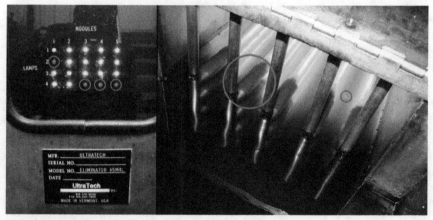

(**Reprinted with permission by Indigo Water Group**)

DATA COLLECTION, SAMPLING, AND ANALYSIS

1. False. Looking directly at a UV lamp may cause conjunctivitis or welder flash.
2. True
3. False. Like a bad sunburn, the effects of welder flash may take several hours to appear.
4. True
5. A
6. C
7. B

Periodic Table

PERIODIC TABLE

group

1	2	3	4	5	6	7	8	9	10	11	12	13	14	15	16	17	18

I A

element name	→	HYDROGEN
atomic number	→	1
chemical Symbol	→	H
atomic weight (u)	→	1,008

VIII A

Alkali metal	Transition metal	Nonmetal	● Solid
Alkaline earth metal	Post-transition metal	Halogen	● Liquid
Lanthanide	Metalloid	Noble gas	● Gas
Actinide			● Unknown

Period 1

HYDROGEN 1 **H** 1,008 — II A — III A IV A V A VI A VII A — HELIUM 2 **He** 4,003

Period 2

LITHIUM 3 **Li** 6,941 — BERYLLIUM 4 **Be** 9,012 — BORON 5 **B** 10,81 — CARBON 6 **C** 12,01 — NITROGEN 7 **N** 14,01 — OXYGEN 8 **O** 16,00 — FLUORINE 9 **F** 19,00 — NEON 10 **Ne** 20,18

Period 3

SODIUM 11 **Na** 22,99 — MAGNESIUM 12 **Mg** 24,31 — III B IV B V B VI B VII B VIII B VIII B VIII B I B II B — ALUMINIUM 13 **Al** 26,98 — SILICON 14 **Si** 28,09 — PHOSPHORUS 15 **P** 30,97 — SULFUR 16 **S** 32,07 — CHLORINE 17 **Cl** 35,45 — ARGON 18 **Ar** 39,95

Period 4

POTASSIUM 19 **K** 39,10 — CALCIUM 20 **Ca** 40,08 — SCANDIUM 21 **Sc** 44,96 — TITANIUM 22 **Ti** 47,87 — VANADIUM 23 **V** 50,94 — CHROMIUM 24 **Cr** 52,00 — MANGANESE 25 **Mn** 54,94 — IRON 26 **Fe** 55,85 — COBALT 27 **Co** 58,93 — NICKEL 28 **Ni** 58,69 — COPPER 29 **Cu** 63,55 — ZINC 30 **Zn** 65,39 — GALLIUM 31 **Ga** 69,72 — GERMANIUM 32 **Ge** 72,59 — ARSENIC 33 **As** 74,92 — SELENIUM 34 **Se** 78,96 — BROMINE 35 **Br** 79,90 — KRYPTON 36 **Kr** 83,80

Period 5

RUBIDIUM 37 **Rb** 85,47 — STRONTIUM 38 **Sr** 87,62 — YTTRIUM 39 **Y** 88,91 — ZIRCONIUM 40 **Zr** 91,22 — NIOBIUM 41 **Nb** 92,91 — MOLYBDENUM 42 **Mo** 95,94 — TECHNETIUM 43 **Tc** (98,91) — RUTHENIUM 44 **Ru** 101,1 — RHODIUM 45 **Rh** 102,9 — PALLADIUM 46 **Pd** 106,4 — SILVER 47 **Ag** 107,9 — CADMIUM 48 **Cd** 112,4 — INDIUM 49 **In** 114,8 — TIN 50 **Sn** 118,7 — ANTIMONY 51 **Sb** 121,8 — TELLURIUM 52 **Te** 127,6 — IODINE 53 **I** 126,9 — XENON 54 **Xe** 131,3

Period 6

CAESIUM 55 **Cs** 132,9 — BARIUM 56 **Ba** 137,3 — LANTHANUM 57 **La** 138,9 — HAFNIUM 72 **Hf** 178,5 — TANTALUM 73 **Ta** 180,9 — TUNGSTEN 74 **W** 183,9 — RHENIUM 75 **Re** 186,2 — OSMIUM 76 **Os** 190,2 — IRIDIUM 77 **Ir** 192,2 — PLATINUM 78 **Pt** 195,1 — GOLD 79 **Au** 197,0 — MERCURY 80 **Hg** 200,6 — THALLIUM 81 **Tl** 204,4 — LEAD 82 **Pb** 207,2 — BISMUTH 83 **Bi** 209,0 — POLONIUM 84 **Po** (210,0) — ASTATINE 85 **At** (210,0) — RADON 86 **Rn** (222,0)

Period 7

FRANCIUM 87 **Fr** (223,0) — RADIUM 88 **Ra** (226,0) — ACTINIUM 89 **Ac** (227,0) — RUTHERFORDIUM 104 **Rf** ---- — DUBNIUM 105 **Db** ---- — SEABORGIUM 106 **Sg** ---- — BOHRIUM 107 **Bh** ---- — HASSIUM 108 **Hs** ---- — COPERNICIUM 112 **Cn** ----

Lanthanides

CERIUM 58 **Ce** 140,1 — PRASEODYMIUM 59 **Pr** 140,9 — NEODYMIUM 60 **Nd** 144,2 — PROMETHIUM 61 **Pm** (144,9) — SAMARIUM 62 **Sm** 150,4 — EUROPIUM 63 **Eu** 152,0 — GADOLINIUM 64 **Gd** 157,3 — TERBIUM 65 **Tb** 158,9 — DYSPROSIUM 66 **Dy** 162,5 — HOLMIUM 67 **Ho** 164,9 — ERBIUM 68 **Er** 167,3 — THULIUM 69 **Tm** 168,9 — YTTERBIUM 70 **Yb** 173,0 — LUTETIUM 71 **Lu** 175,0

Actinides

THORIUM 90 **Th** (232,0) — PROTACTINIUM 91 **Pa** (231,0) — URANIUM 92 **U** (238,0) — NEPTUNIUM 93 **Np** (237,0) — PLUTONIUM 94 **Pu** (239,1) — AMERICIUM 95 **Am** (243,1) — CURIUM 96 **Cm** (247,1) — BERKELIUM 97 **Bk** (247,1) — CALIFORNIUM 98 **Cf** (252,1) — EINSTEINIUM 99 **Es** (252,1) — FERMIUM 100 **Fm** (257,1) — MENDELEVIUM 101 **Md** (256,1) — NOBELIUM 102 **No** (259,1) — LAWRENCIUM 103 **Lr** (260,1)